New Mexico Magazine
nmmagazine.com

Author/Photographer: Steven J. Cary
Editors: Penny Landay, Emily Drabanski
Design & Production: Bette Brodsky
Maps & Illustrations: Bette Brodsky

Library of Congress PCN Number: 2009925231
ISBN: 978-1-934480-03-8
Printed in Korea

Cover photos:

Dog Canyon, Oliver Lee Memorial State Park, in the Sacramento Mountains

Butterflies, clockwise from upper left:

RED ADMIRAL
TWO-TAILED SWALLOWTAIL
VARIEGATED FRITILLARY
GIANT SWALLOWTAIL

Previous page: GRAY CRACKER

Opposite page, left to right:

ISABELLA'S HELICONIAN
MEXICAN FRITILLARY

Back cover photo: SOUTHERN DOGFACE

BUTTERFLY LANDSCAPES *of* NEW MEXICO

Steven J. Cary

NEW MEXICO
MAGAZINE

CONTENTS

Introduction 6

How to Use this Book 7

ABOUT BUTTERFLIES 8

- Colorful Wings 9
- Incredible Life Cycle 10
- Appreciating Butterflies 12
- Butterfly Gardening 14

BUTTERFLY LANDSCAPES 16

- Here, There, Everywhere 17
- Vertical Parade 26
  - Upper Sonoran Zone 30
  - Transition Zone 39
  - Canadian Zone 52
- Regional Specialties 61
  - Eastern Plains 64

From top:

SILVER-BANDED HAIRSTREAK
WHITE-PATCHED SKIPPER

North-Central Mountains 76
Northwest Plateau 94
Southwest Basin and Range 106
Desert Borderlands 120

ADDITIONAL RESOURCES 138
Accessing Butterfly Landscapes 139
Checklist of New Mexico Butterflies 140
Butterfly Glossary 146
Recommended Reading 148
State Parks 150
State Map 151
Index 152
ACKNOWLEDGMENTS 166
PHOTO CREDITS 167

From top:

DRUSIUS CLOUDYWING
RUSTY-TIPPED PAGE

INTRODUCTION

Butterflies have always brought magic to my life. I fell in love with these gossamer-winged creatures when I was growing up in Wisconsin. You can imagine my delight when I moved more than 25 years ago to the Land of Enchantment—a virtual butterfly paradise. New Mexico has more than 300 different kinds of butterflies; only Texas and Arizona have more.

I collected butterflies as part of my initial exploration of the natural world. For most of my adult life I've captured their images with camera or binoculars and taken extensive field notes. After exploring many of New Mexico's nooks and crannies, I've amassed much information about our butterflies. As New Mexico State Parks' chief naturalist, I enjoy talking with countless and varied park visitors—wide-eyed children curious about nature, amateur butterfly watchers, and scientists seeking to learn about the secret, yet vital, world of insects. Seeing a new butterfly for the first time still brings me the same excitement and joy I felt as a youth.

Many excellent butterfly field guides are available and are helpful in identifying New Mexico's butterflies. I, however, wanted to go beyond the rigid guidebook format and gather my information into a book that could enhance your understanding of and delight in these spectacular butterflies. I'm a geographer at heart, so *Butterfly Landscapes of New Mexico* reveals the state's butterflies not merely as beautiful animals but also as complex expressions of local geology, landforms, climate, and plants.

Every major section of the book begins with landscapes and habitats; the text that follows explains which butterflies live there and illuminates their roles in those places. For each part of the state, butterflies are first portrayed within a large region, then in progressively smaller components of landscape—from mountain ranges to valleys to hilltops.

Exploration of these areas highlights butterflies that are distinctive to particular seasons of the year, unusual plant communities, or specialized habitats nested within a region.

Enjoy this journey into the world of these charismatic insects, dwell in their habitats and landscapes, savor their beauty, and discover their contributions to New Mexico's diverse ecological settings.

—**Steven J. Cary**

GIANT WHITE

HOW to USE this BOOK

Butterfly Landscapes of New Mexico is a geographic and photographic guide to the remarkable array of butterflies that grace our state. Useful at home or in the field, it provides a strong introduction to the 300-plus species of butterflies reported from New Mexico. But the subject is wide and deep, and readers may wish to use this book in tandem with other pertinent field guides.

Butterfly Landscapes presents New Mexico's butterflies in three sections: (1) ubiquitous species seen almost anytime, anyplace; (2) species typical of prevailing elevations or life zones; and (3) specialized species in different geographic areas. Butterflies identified with blue captions and scattered throughout are subtropical strays that have wandered in from Mexico only once (that we know of).

NAMES. Scientists studying butterflies continually generate new ideas about relationships between different kinds, which often results in new scientific names. Common names are no different. To present butterflies as accurately as possible, the author used the latest authorities. Common names are given first for each species; they follow the North American Butterfly Association (NABA) checklist as closely as possible. Scientific (Latin) names consistent with the 2008 J. P. Pelham catalogue follow in italics and brackets. Host plants' scientific names also appear in italics and brackets.

ARRANGEMENT. Most field guides arrange butterflies to facilitate identification or to indicate genetic relationships. This book, however, organizes butterflies according to how and where they are found in nature.

PHOTOS. Color photographs help readers recognize butterflies they observe. The author captured the majority of the book's images during almost three decades of chasing butterflies in New Mexico. Other photographers are identified when their photos are used (see page 167). Butterflies are shown at approximately life size.

SPECIAL ELEMENTS. Sidebars inserted throughout the text illuminate some of the intriguing stories behind the main story. *History Highlights* tell about important people, times, and circumstances in New Mexico's butterfly past. *Fine Dining* sidebars describe certain plants or plant groups that are popular as caterpillar food and thus contribute a little extra to local butterfly landscapes. In addition to being cultural icons, butterflies are vital cogs in local ecosystems. *Eco-Notes* reveal some of the crucial roles that butterflies play as munchers of plants, pollinators of flowers, partners to ants, or food for predators.

TIP: *See the Additional Resources section for a recommended reading list, glossary, butterfly checklist, and other useful information.*

ABOUT BUTTERFLIES

- COLORFUL WINGS
- INCREDIBLE LIFE CYCLE
- APPRECIATING BUTTERFLIES
- BUTTERFLY GARDENING

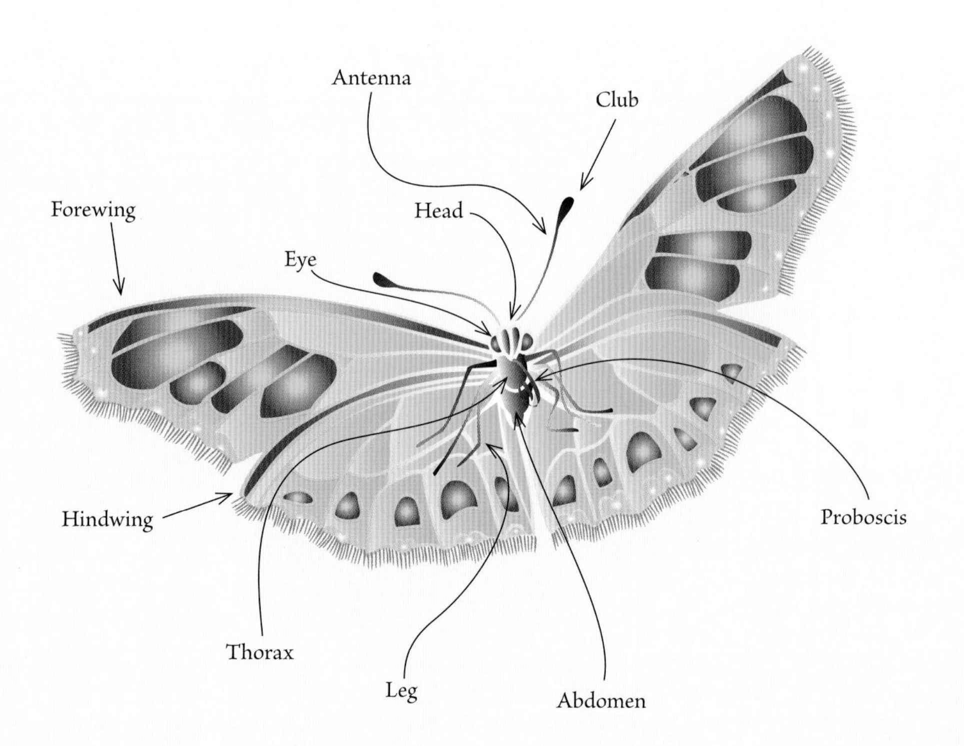

Butterflies are ambassadors to modern civilization from an obscure and ancient world. With their ephemeral beauty and captivating grace, they lead observers through a portal into the world of insects. Insects are the most diverse group of animals on Earth and represent more than half of all known living organisms.

Butterflies and other insects are not like us. Humans have backbones and internal bony skeletons, while insects lack a spine and are supported by an external skeleton made of chitin, a tough, protective material. Humans generate their own body heat, but butterflies must gather heat from their surroundings to be active.

Butterfly bodies are divided into three principal parts: head, thorax, and abdomen. The head has two compound eyes, two clubbed antennae that gather olfactory information, and a tube-like mouth called a proboscis. Digestive and reproductive systems are housed in the abdomen. Adult butterflies have six legs and four wings, all powered by muscles in the thorax.

Most people experience butterflies as the brightly colored, actively flying adults. Color resides in scales laid on the wings like thousands of shingles on a tiny roof. Each scale has its individual color. Yellows, oranges, browns, and whites are usually derived from pigments inside the scales. Some scales also have surfaces that refract light to produce iridescent greens, blues, and true reds. Scale-winged insects—butterflies and moths—are collectively known as the Lepidoptera.

Humans are visual creatures, so we appreciate the splendor of brilliant wings, but scaled Lepidoptera wings also serve essential functions for their owners. The powerful desire to avoid being eaten long enough to find a mate influences the animals' wing shapes, colors, and patterns. Most moths fly at night when many predators cannot see to hunt and then camouflage themselves during their daytime rest. Butterflies, in contrast, have conquered daylight. They have the tools to be active when visual predators are active: good vision and multicolored wings.

In addition to powering flight, butterfly wings also camouflage their owners and attract mates. Wings have upper and lower surfaces that can be colored, patterned, and wielded separately for these seemingly contradictory purposes. The brightest colors usually are on the upper sides; butterflies show these surfaces when flying or when basking to warm themselves in the morning sun. Upper wing surfaces display color patterns that allow other butterflies to determine if the creatures belong to their species. Males of some kinds also shed scented scales (cologne!) to convince females they are worthy mates, but colorful wings provide the primary enticement.

All butterflies close their wings when they are inactive—at night, during cool or wet weather, when it is too cold to fly, and, for some, during hibernation. For prey animals such as butterflies, it is best to be invisible while inactive. If spotted while sleeping, they will be munched before dawn. Most butterflies have wing undersides that make them hard to see, or at least hard to recognize as something to eat. Some blend in with green vegetation where they rest. Others have wings patterned to mimic more threatening animals like spiders or small mammals. Butterflies that hibernate as adults sport wing colors, patterns, and shapes resembling clumps of dead leaves or loose tree bark where they spend winter. Some butterflies have underside patterns and shapes that trick birds into attacking disposable parts of their wings, allowing escape. The protective disguises are diverse and amazing.

INCREDIBLE LIFE CYCLE: KEY to BUTTERFLY LANDSCAPES

We can easily appreciate the beauty of adult butterflies, but the adult stage is just the most visible form of a complex creature. The less visible parts determine how butterflies arrange themselves in New Mexico's spectacular landscape.

Butterflies undergo an extraordinary life cycle that involves complete metamorphosis from immature stages to adulthood. After mating, each adult female places a few hundred eggs throughout her environment. She maximizes survival of her offspring by providing them with the best possible food. How? She places her eggs on plants that her young, upon hatching, will happily eat.

A butterfly egg normally matures in a few days, after which the larva, or caterpillar, chews its way out. The caterpillar is the growing stage of a butterfly's life cycle. Larvae eat plant material, which can include stems, leaves, roots, flowers, or seeds, depending on the butterfly species. Growing poses challenges for caterpillars because their skeletons are on the outside (exoskeletons). They grow in stages called instars; after a young caterpillar fills its first, tiny exoskeleton, it sheds that skin, creates a larger version, and resumes feeding. Relentlessly eating and excreting, it increases its body mass about 10,000 times between first and last instars.

Upon filling its final larval exoskeleton, a caterpillar finds an appropriate place to create a new skin—the chrysalis (or pupa). Metamorphosis occurs inside this shell. The organism begins its transformation as a worm-like creature—flightless, plant chewing, asexual, and with limited sensory abilities. Then magic happens. After a few weeks the chrysalis turns transparent and reveals the adult butterfly, ready to emerge.

The same individual climbs out—on the other side of adolescence—fully grown. Adults have sophisticated legs and wings: they can fly! They have complex eyes and see very well. Antennae and feet house olfactory senses, but North American butterflies cannot hear. In place of the mandibles that caterpillars used to chew solid food, each adult has a flexible straw-like tube (proboscis) to siphon liquids. The primary mission during the brief two weeks (more or less) of adulthood is to mate and reproduce.

All butterflies experience this sequence, but each species adjusts the details to maximize survival. The timing of adult flight varies among butterfly species; in every case the life cycle is timed so larvae can eat the host plant when it is the most nutritious. Winter is spent in whatever stage supports the timing requirements of larvae. Some species spend winter as eggs, others as larvae, some as chrysalids, and a few as adults. Butterflies occur where their host plants occur; those plants grow where conditions suit them; and each butterfly's menu of caterpillar plant food links that species to its special niche in the landscape.

LIFE CYCLE *of a* BUTTERFLY

APPRECIATING BUTTERFLIES

Until the late twentieth century, the most popular way to appreciate butterflies was to collect them. Collections were the basis for the scientific work that made this and other books possible. Before 1980, butterfly field guides were illustrated almost exclusively with drawings, paintings, or photos of pinned specimens. Responsible collecting remains important today as a foundation for scientific research. Young people also find collecting to be a tactile and rewarding learning experience.

A modest number of butterfly specimens usually can be collected with little impact on wild populations. A properly pinned, spread, and labeled specimen is still the most reliable way to document an observation. Equipment and supplies for collecting and caring for specimens are affordable and widely available. Collecting for personal gain, however, is to be avoided, particularly where natural landscapes are disappearing and native butterfly populations are declining.

Since the 1970s, technological advances have boosted the popularity of other forms of butterfly observation and study. Affordable high-quality color macrophotography is now available. Photos from nature contain much behavioral information that is lost when a specimen is collected. Digital cameras are replacing film cameras, with no sacrifice in quality, and resulting images can be electronically transmitted around the world. Color macrovideography adds the valuable dimension of motion. Close-focus binoculars bring butterflies into better view for butterfly watchers.

Whatever your approach to butterfly observation, it's best to keep records of what you see. Document your sightings with a well-kept field notebook, properly labeled specimens, and photographs or accurate drawings. Include in your notes the relevant time, place, elevation, habitat, and weather conditions, in case you want to find a particular critter again in a few years. Noting where females place their eggs, which caterpillars you find on which plants, and what occurs during growth stages will enhance your understanding of butterfly life histories. It's always fun to look back at notations from earlier outings so you can recall what you saw or share your discoveries with others.

Right:
BLACKENED BLUEWING
Opposite page:
MEXICAN SILVERSPOT

WATCHING:
Camera
Field Notebook
Binoculars
Field Guides

Barbara Diener

COLLECTING:
Net
Spreading Board
Forceps

TIP: *For more information about collection and observation techniques see the listings under Methods and Tools in the Recommended Reading section on page 149.*

BUTTERFLY GARDENING

Butterfly gardening provides a wonderful way to learn about and interact with butterflies. It's like stocking a bird feeder, because the type of feeder and food you offer influences the birds that come to visit. But compared to birds, butterflies are less predictable in occurrence and subtler in their behaviors. Because of the creatures' complex life cycles, butterfly gardening has two aspects: planting nectar sources for adults and planting food for caterpillars.

Attracting adults to the garden brings faster results. Most adult butterflies are opportunistic feeders: they will readily come to flowers that offer nectar within reach of their proboscides. Not all flowers produce nectar, however, and a good first step is to find out which plants are popular among butterflies in your area. Most hungry butterflies favor large clusters of small flowers. Beebalm [*Monarda*], verbena [*Glandularia*], dogbane [*Apocynum*], milkweeds [*Asclepias*], and plants in the aster or daisy family (Asteraceae) use that strategy. If a butterfly has success at one flower, it instinctively looks for more of the same because that's safer than flying around, dodging predators, and gambling on something different. If you patronize big-box stores you know what I mean. It's better to plant large numbers of a few kinds of flowers than one each of many kinds. Your nectar garden will provide a window into the fascinating world of pollination ecology.

Adult butterflies also need water and salts, which in nature they get from moist soil, animal scat, and rotting matter. Birds can drink from open water sources and enjoy birdbaths, but butterflies can drink only by siphoning liquids from porous material such as soil. To attract butterflies, make a sunny, wet spot in your butterfly garden and occasionally add salt and spoiled fruit, the stinkier the better.

A more challenging way to garden for butterflies is to provide food for caterpillars. Study a few butterflies that are routine in your area, learn what their caterpillars eat, and then plant those plants—lots of them. Gardeners have been taught to pull off every caterpillar they see, but please resist that urge. Don't use pesticides or insecticides either. Gardening for caterpillars is demanding on several levels, takes more time, and makes you see plants differently, but the rewards are proportionally greater.

The use of native plants in all gardens is crucial. Non-native plants, like butterfly bush [*Buddleia*], are very attractive to butterflies, but if the critters spend their feeding time at non-natives, how will our native plants be pollinated? Many native ecosystems are under assault from invasive exotic plants, so gardeners should support indigenous species that help make New Mexico different from other places—enchanting even!

TIP: *Local chapters of the New Mexico Native Plant Society are excellent resources for native plants. For more information on butterfly gardening see the listings under Gardening in the Recommended Reading section on page 149.*

SILVER EMPEROR

BUTTERFLY LANDSCAPES

- **HERE, THERE, EVERYWHERE**
- **VERTICAL PARADE**
  - Upper Sonoran Zone
  - Transition Zone
  - Canadian Zone
- **REGIONAL SPECIALTIES**
  - Eastern Plains
  - North-Central Mountains
  - Northwest Plateau
  - Southwest Basin and Range
  - Desert Borderlands

The first step in exploring New Mexico's butterflies and butterfly landscapes is an easy one. No matter where you find yourself in New Mexico, you can see many familiar friends simply by stepping outside on sunny summer days.

Some 28 different kinds of butterflies are essentially ubiquitous in the state. They are indifferent to subtleties of habitat or landscape—equally content on Wheeler Peak, the Llano Estacado, or urban alleys. They won't all be there the first time you look, or at any other single time, for that matter. But they will gradually reveal themselves over weeks and months as you keep a sharp eye out in your neighborhood.

Butterflies naturally seek out plants that offer the finest dining opportunities for their offspring. One butterfly species has larvae that eat almost any plant. GRAY HAIRSTREAK [*Strymon melinus*] caterpillars are happy to munch on plants in more than 60 different plant families. With such an undiscriminating palate, this species finds a toehold virtually everywhere.

Top right:
GRAY HAIRSTREAK

Above to below:
CABBAGE WHITE
ORANGE SULPHUR

Three species like to frequent farms and gardens. CABBAGE WHITE [*Pieris rapae*], the state's only resident European immigrant, thrives in urban gardens where people plant cabbage, kale, broccoli, ornamental mustards, or other non-native plants in the mustard family. You may even find it in long-abandoned ghost towns. Strangely, its larvae will not eat the numerous native mustards. CLOUDED SULPHUR [*Colias philodice*] and ORANGE SULPHUR [*Colias eurytheme*] have caterpillars that eat all legumes and are especially fond of alfalfa. You can see them by the thousands near alfalfa fields around Artesia in late summer, but they are routine almost everywhere.

Left to right:
CLOUDED SULPHUR
DAINTY SULPHUR

Legumes, any old legumes at all, also host MARINE BLUE [*Leptotes marina*] and REAKIRT'S BLUE [*Echinargus isola*]. The two winged sapphires are tiny; you have to look very closely (or use close-focus binoculars) to see the identifying marks. But these blues are prolific and widespread. MELISSA BLUE [*Plebejus melissa*] belongs in this group as well, although the females are somewhat pickier about the legumes they select for caterpillar food.

Some resident native butterflies follow landscape disturbances. Prominent among these is CHECKERED WHITE [*Pontia protodice*], whose caterpillars eat native mustards, including prevalent tansy mustard. DAINTY SULPHUR [*Nathalis iole*] larvae eat roadside plants in the composite (daisy) family. Wherever people disturb land

Above to below:

REAKIRT'S BLUE
MARINE BLUE

Right and top:

CHECKERED WHITE
MELISSA BLUE

for roads, subdivisions, agriculture, or even by accident, various pioneer plants that colonize such areas will host these butterflies. COMMON CHECKERED-SKIPPER [*Pyrgus communis*] and its twin, the WHITE CHECKERED-SKIPPER [*Pyrgus albescens*], are in this group, too, though each is not truly ubiquitous. The former lives in northern New Mexico and high altitudes while the latter prevails in southern New Mexico and low elevations. They are indistinguishable to the naked eye, however, and occupy similar habitats. PAINTED CRESCENT [*Phyciodes picta*] thrives on weedy asters and the widespread field bindweed [*Convolvulus arvensis*].

Above, top to bottom:

WHITE CHECKERED-SKIPPER
PAINTED CRESCENT

Bottom, left to right:

COMMON CHECKERED-SKIPPER
RED ADMIRAL

Above:
MOURNING CLOAK
Right:
AMERICAN LADY
Below:
VARIEGATED FRITILLARY

Several butterflies are pickier eaters but, luckily for them, have host plants that are plentiful. This group includes VARIEGATED FRITILLARY [*Euptoieta claudia*], MOURNING CLOAK [*Nymphalis antiopa*], RED ADMIRAL [*Vanessa atalanta*], and AMERICAN LADY [*Vanessa virginiensis*].

Top row, left to right:
QUEEN
FUNEREAL DUSKYWING
SACHEM
Below:
MONARCH

Some butterflies are common residents only in southern New Mexico, but their traveling habits take them regularly throughout the state. This group includes PIPEVINE SWALLOWTAIL [*Battus philenor*], SACHEM [*Atalopedes campestris*], AMERICAN SNOUT [*Libytheana carinenta*], COMMON BUCKEYE [*Junonia coenia*], FUNEREAL DUSKYWING [*Erynnis funeralis*], and QUEEN [*Danaus gilippus*].

Not every butterfly can survive winters here. A few species get frozen out but then ride favorable winds back to New Mexico the following warm season. They breed throughout that time with varying degrees of success. PAINTED LADY [*Vanessa cardui*] and MONARCH

[*Danaus plexippus*] are the best known of this group. Each migrates north from Mexico in springtime, sometimes in large numbers. Only Monarchs make a return flight in autumn, usually in September and October.

Above to below:

PAINTED LADY

AMERICAN SNOUT

Right:

COMMON BUCKEYE

Below:

PIPEVINE SWALLOWTAIL

New Mexico's other regular summer vacationers are subtropical sulphurs: SLEEPY ORANGE [*Abaeis nicippe*], SOUTHERN DOGFACE [*Zerene cesonia*], CLOUDLESS SULPHUR [*Phoebis sennae*], and MEXICAN YELLOW [*Eurema mexicana*]. Most use legumes as caterpillar food and can complete one or two generations during the warm season. Winters here are too cold for them, regardless of life stage, but another generation heads north from Mexico each spring. If global warming causes milder winters, look for these sulphurs to establish themselves as year-round residents in southern New Mexico.

Right, top to bottom:
SLEEPY ORANGE
MEXICAN YELLOW
CLOUDLESS SULPHUR
Below:
SOUTHERN DOGFACE

Life Strategies

Why is it that some butterflies are everywhere, but to see others you have to mount carefully planned expeditions to out-of-the-way places? The answer lies in the plants each species needs in order to survive. Butterflies tend to adopt life strategies similar to the life strategies of the plants their caterpillars eat.

Plants that occupy shady, wooded habitats tend to grow slowly, produce few seeds, and disperse them poorly. They tend to be large perennials (trees or shrubs) rather than small annuals. Butterflies whose larvae eat these plants have compatible characteristics: slow-growing larvae, one generation per year, very narrow larval food choices, and adults that rarely wander far from stands of the host. Considerable detective work may be necessary to find these varieties.

In contrast, plants that thrive in often-disturbed, sunny habitats tend to grow fast, flower early, produce many seeds quickly, and disperse them widely. Butterflies whose larvae eat these plants tend to have similar traits: fast-growing caterpillars, two or more generations per year, broad host-plant options, and active adults flying in large, wide-ranging populations.

Butterflies that we see more frequently, such as Painted Lady, Checkered White, Variegated Fritillary, Common Checkered-Skipper, and Dainty Sulphur, have large populations of actively wandering adults. Their females place eggs on plants that are widely distributed and thrive in disturbed areas (O.K., weeds!). They breed continuously as long as weather permits, and we see these butterflies (flying weeds!) almost anywhere, anytime.

VERTICAL PARADE

When you become familiar with the state's 28 common butterfly species, your next step is to venture into natural habitats. You may have to move slightly farther from your back door, but the sojourn need not exceed one to two hours each way. For this modest investment of time you can expect a reward of an additional 60-plus butterfly species, tripling the number of species in your field notebook. To organize this effort, you may wish to learn more about New Mexico's life zones.

The state's span of elevations from roughly 3,000 feet to 13,000 feet contributes significantly to its diversity of butterfly landscapes. Across this range, plants and animals occur in five biological life zones—from Upper Sonoran through Transition, Canadian, Hudsonian, and Arctic. The state is home to lowland desert butterflies, Arctic tundra butterflies, and everything in between. It lacks only Lower Sonoran, Subtropical, and Tropical life zones.

Life zones represent a classification of lands according to length of growing season, which is longest at low elevations and shortest at high elevations. Each life zone has its characteristic vegetation, which may include trees as well as grasses, forbs, shrubs, and other nonarboreal plants. Similarly, each zone has wet regions with their associated plants and animals. Mountains have a way of compressing into a small area the same life zones that otherwise would require a trip from Texas to the Arctic Circle.

As with most classifications, you can expect considerable blurring of boundaries between life zones. For example, if you walk around any major hill or mountain while staying at the same elevation, you usually find plants and animals typical of a higher life zone on the north-facing slope in contrast to what's found on the south-facing slope.

Most New Mexico lands lie within the Upper Sonoran, Transition, or Canadian zones. The following three sections delve into these "big three" life zones and explore their butterfly faunas. In this stage of your inquiry, the focus of your travels will be the central portions of New Mexico. The Río Grande Valley is, in many respects, the geographic heart of the state. It unifies much of New Mexico from a drainage perspective, while dividing it biogeographically. Stretching from Colorado to Texas and from 3,300 feet to 10,000 feet in adjacent uplands, it offers a representative sample of the three life zones.

New Mexico has more extreme elevations and associated life zones, but they are restricted to small areas and have fewer butterflies—though often these species can be seen nowhere else in the state. You can read about these regional specialties in subsequent sections.

LIFE ZONES
ARCTIC
tundra
12,000' to 13,000'
25" to 28" annual precipitation
HUDSONIAN
spruce/fir
10,000' to 12,000'
21" to 23" annual precipitation
CANADIAN
mixed conifer
8,000' to 10,000'
17" to 19" annual precipitation
TRANSITION
pine/oak/juniper
6,000' to 8,000'
15" to 17" annual precipitation
UPPER SONORAN
grassland/scrub
3,000' to 6,000'
8" to 15" annual precipitation

Where Are the Butterflies?

ECO-NOTES

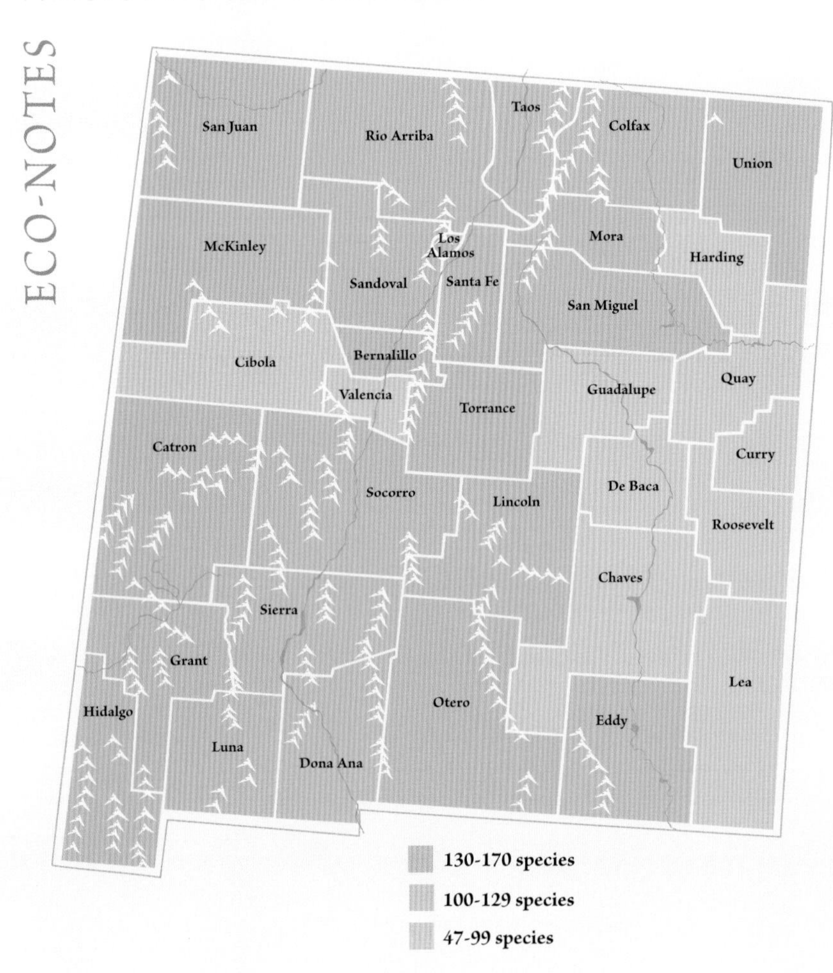

Butterflies are not distributed equally or uniformly throughout New Mexico, either in terms of absolute numbers or species diversity.

Butterfly diversity is greatest in areas of greatest plant diversity. Plant diversity is greatest in areas of greatest topographic relief because such areas offer the greatest number of ecological niches for different plants to grow. Hence, you might want to head to the mountains to find the most varieties of butterflies.

This map shows the general number of butterfly species found (so far) in each of New Mexico's 33 counties.

HISTORY HIGHLIGHT

Charles Townsend

In 1889 the New Mexico College of Agricultural and Mechanical Arts, now New Mexico State University, was established at Mesilla. Entomologist Charles Henry Tyler Townsend joined the faculty in 1891; we owe much of our understanding of life zones to him.

Townsend was hired to educate students and assist farmers in dealing with crop pests. Upon arrival, he immediately began to teach classes and study local insects, many of which were as yet unknown to science. His early findings were published in the college's *Experiment Station Bulletins*. Although Townsend left his university position in 1893, he frequently returned to New Mexico and the Southwest to pursue his passion for flies (yes, Diptera!) and biogeography.

Townsend's primary study areas were near Las Cruces. But a naturalist in a new territory is like a kid in a candy store, and Townsend wanted to explore different locales and find new flies. In 1891 he traveled across western New Mexico to Flagstaff, Arizona, and the south rim of the Grand Canyon, collecting about 1,100 specimens along the way. In July 1894 he made an expedition to the headwaters of the Gila River. He was probably the first naturalist to do so and to report findings in peer-reviewed journals. He also made a collecting expedition into the Sacramento Mountains in 1896.

Concepts of biogeography and life zones were the big buzz among naturalists in the 1890s. Scientific discourse began in 1893 when C. Hart Merriam published a map delineating life zones for the United States as a whole. On the basis of his own travels in Arizona, West Texas, and New Mexico, Townsend refined Merriam's proposal. In 1895 he published a comprehensive regional analysis of vegetation and life zones in the American Southwest. He identified five life zones and applied them to areas he knew in New Mexico including the Gila region; Mount Taylor; and the San Andres, Sacramento, Florida, and Sandía mountains. Townsend's system included, from low to high elevation, the Upper Sonoran, Transition, Canadian, Hudsonian, and Arctic zones and is still used today.

Clockwise from top left:

BRONZE ROADSIDE-SKIPPER
VIERECK'S SKIPPER
OSLAR'S ROADSIDE-SKIPPER

Begin your exploration in the Upper Sonoran Zone, which ranges from roughly 3,000 feet to 6,000 feet in elevation. This foray into the great New Mexico butterflyscape explores gently sloping grasslands, often with interspersed oak, juniper, or desert scrub. If you hear meadowlarks and feel the wind blowing, you are in the right place. This swath of habitats is widespread and occurs in every New Mexico county. I-40 crosses the Upper Sonoran Zone from Tucumcari to Santa Rosa and from Albuquerque to Grants. I-25 traverses it from Socorro to Santa Fe and from Las Vegas to Raton. I-10 crosses it from Las Cruces to Lordsburg. The communities of Clayton, Fort Sumner, Roswell, Carlsbad, Albuquerque, Farmington, Carrizozo, and Las Vegas are located in this life zone.

Butterflies typical of the Upper Sonoran Zone are viewable statewide, more or less, but their habitats are optimal in just a few places. These areas include eastern and southern New Mexico's rolling grasslands, canyon/mesa country, and low mountain foothills. Mid-May is a good time to visit because weather is consistently warm, days are long, and thistles are in bloom. Thistles are magnets for species with the elongated feeding apparatus necessary to reach nectar at the base of the plants' flowers.

Gully bottoms are often productive places because that's where males look for females. Males will take time from their mate searches to gather nectar from irresistible thistles. On a good May day at Oliver Lee Memorial State Park, south of Alamogordo, you can see GREEN SKIPPER [*Hesperia viridis*] and VIERECK'S SKIPPER [*Atrytonopsis vierecki*] elbowing each other for access to the best thistles. When in the caterpillar form, each eats grasses. Green and Viereck's were unknown to science until discovered, respectively, in 1882 by Francis Snow near Las Vegas and in 1902 by Henry Viereck near Alamogordo. When you spot these skippers you can recall this tidbit of New Mexico butterfly history.

GREEN SKIPPER

In July, after the summer rains begin, journey to Conchas Lake State Park to seek more grass-feeding skippers that typify Upper Sonoran butterfly habitats. Visit gullies to find OSLAR'S ROADSIDE-SKIPPER [*Amblyscirtes oslari*] and BRONZE ROADSIDE-SKIPPER [*Amblyscirtes aenus*]. Bronze prefers rocky canyon bottoms, while Oslar's prefers gentler, grassier prairie swales.

Then climb a low hill to find PAHASKA SKIPPER [*Hesperia pahaska*]. Males fly to local hilltops where they perch on the ground and chase all comers with pugnacity typical of much larger creatures. To replenish energy resources, adults seek out purple nectar. FULVIA CHECKERSPOT [*Chlosyne fulvia*] flies here, too. You'll find the hills relatively easy to climb at Cerrillos Hills State Park in Santa Fe County.

Many Upper Sonoran landscapes consist of rolling or slightly rough country with shallow drainages in the form of creeks or arroyos. Rough or rocky slopes and hills of varying ruggedness often border these temporary watercourses. This arrangement of varied landscapes has excellent plant diversity, packs several habitats into a small area, and offers good opportunities for butterflyers willing to explore hilltops, gully bottoms, and everything in between. The frequent juniper shrubs in this type of country are hosts for JUNIPER HAIRSTREAK [*Callophrys gryneus*].

Look in the creek beds for puddles or wet patches of sand or mud. You can find almost anything at these great butterfly magnets, especially if the weather is hot and dry. Walk slowly, crisscross the bottom, and scan for slight movements. You might see two of the Southwest's tiniest butterflies, ACMON BLUE [*Plebejus acmon*] and

SMALL CHECKERED-SKIPPER [*Pyrgus scriptura*], which are encountered irregularly through the warm season.

While walking head-down in an arroyo, you may be surprised by the shadow of our largest butterfly. With a wingspan approaching six inches, the TWO-TAILED SWALLOWTAIL [*Papilio multicaudata*] flies from April through August, often soaring majestically along canyons such as at Percha Creek near Hillsboro. It also feeds at thistles and wet sand. Though seen in much of New Mexico, it is most common in the southern areas. Its caterpillars eat the foliage of hoptree [*Ptelea trifoliata*], chokecherry [*Prunus virginiana*], and velvet ash [*Fraxinus velutina*]. City gardens attract this butterfly, and urban ash plantings allow it to establish breeding populations in Albuquerque and Santa Fe.

Opposite page, top to bottom:

JUNIPER HAIRSTREAK
FULVIA CHECKERSPOT
PAHASKA SKIPPER
ACMON BLUE

Right:

SMALL CHECKERED-SKIPPER
TWO-TAILED SWALLOWTAIL

SPRING WHITE

Spring is prime time for butterflying in New Mexico's low elevations. The foothills of the Gallinas Mountains west of Corona provide havens for hilltopping butterflies in search of mates. Examples are SLEEPY DUSKYWING [*Erynnis brizo*], BLACK SWALLOWTAIL [*Papilio polyxenes*], and YUCCA GIANT-SKIPPER [*Megathymus yuccae*].

The spring hilltopping crowd also includes SPRING WHITE [*Pontia sisymbrii*] and SARA ORANGETIP [*Anthocharis sara*]. Each selects mustards for its caterpillars to eat. Mustards sprout, bloom, and go to seed by late spring. By June all larvae have fed, pupated, and entered a diapause that lasts at least until the following spring.

Left to right:

YUCCA GIANT-SKIPPER
SLEEPY DUSKYWING
SANDIA HAIRSTREAK

Between creek beds and hilltops you will often find grasslands or rocky terrain. South-facing slopes tend to be on the warm, dry side. If you come upon a big patch of Texas beargrass [*Nolina texana*] in the spring, look there for SANDIA HAIRSTREAK [*Callophrys mcfarlandi*]. Small green-and-gold adults fly in greatest numbers in April. They perch on beargrass leaves and nectar nearby. This species was described in 1959 from specimens taken in the Albuquerque foothills of the Sandía Mountains; La Cueva and Elena Gallegos recreation areas remain great places to look. New Mexico's Wild Friends successfully sponsored legislation in 2003 to make Sandía Hairstreak the official state butterfly.

In similar foothill settings, such as the Silver City area, visit patches of wild buckwheats [*Eriogonum wrightii*] in late August to glimpse RITA BLUE [*Euphilotes rita*]. These tiny, colorful blues perch on their hosts and seek nectar and moisture nearby.

BLACK SWALLOWTAIL

Bottom, left to right:
RITA BLUE
SARA ORANGETIP

Clockwise from above:
WESTERN PYGMY-BLUE
COMMON SOOTYWING
SALTBUSH SOOTYWING

Some Upper Sonoran habitats in New Mexico are a tad on the salty side. This condition may result from naturally occurring gypsum in local rocks and soils or long-term evaporation in low-lying areas that collect seasonal water. These locales are usually occupied by salt-tolerant plants, which in turn support their cadre of herbivorous butterflies. Widespread application of salt to de-ice winter roads often extends the habitats into otherwise non-salty areas.

Three butterflies have caterpillars that prefer salt-tolerant plants. Look for these small, weakly flying creatures at Bottomless Lakes State Park and White Sands National Monument. SALTBUSH SOOTYWING [*Hesperopsis alpheus*] flies from April through May and again in July. Four-wing saltbush [*Atriplex canescens*] is the main host here, as it is for WESTERN PYGMY-BLUE [*Brephidium exilis*], the state's smallest butterfly. Adults are found year-round in the south, and summer wanderers to the north may produce offspring before winter arrives. COMMON SOOTYWING [*Pholisora catullus*] inhabits disturbed and saline areas almost statewide below 7,000 feet. Its larvae eat various goosefoots [*Chenopodiaceae*] and amaranths [*Amaranthaceae*]. Adults are most prevalent from June to September.

Once There, Where Do You Look? Or What Is Habitat?

Habitat for any butterfly is the place where it can find appropriate food, water, sun, and shelter. Ignoring for now the caterpillars, for which habitat is essentially the larval host plant, let's examine habitat needs of adult butterflies. Understand these and you'll know where to look for butterflies within the landscape.

Adults of most butterfly species live long enough and are sufficiently active to require supplemental energy, which they obtain from food. The adult feeding apparatus is the proboscis, so solid food is not an option. Ecosystems provide a surprising abundance and diversity of liquids containing energy, vitamins, and minerals. Flower nectar is the most popular source of energy and nutrition for adult butterflies, but some prefer tree sap or rotting fruit.

The creatures also need water, as well as dissolved salt and other minerals, to maintain proper physiological function. Water sources are particularly important in semi-arid New Mexico, especially in times of drought or on hot summer days. That's when adult butterflies often sit at puddle edges and siphon water from saturated soil.

Butterflies require sun and warmth to be active. Adults exist to mate; they can't find mates unless they can fly around; and butterfly wing muscles need to maintain a temperature of about 90°F in order to function. Not surprisingly, most butterflies prefer sunny places. After cool New Mexico nights, their first morning task is to bask in the sun to raise their body temperature.

Shelter is a final key element of butterfly habitat. Butterflies necessarily have a lot of quiet time—at night or in cool weather. When immobile, they require protection from normal environmental hazards including wind, rain, snow, hail, and predators. Shelter takes the form of brush, foliage, crevices in rocks, or loose tree bark. Adults that hibernate over winter require superior crannies for their six-month snooze.

Blue Grama Grass

FINE DINING

Blue grama grass [*Bouteloua gracilis*] is one of the most important plants of the high plains. It dominates semi-arid grasslands from Mexico to Wyoming, and its nutritious parts that once served millions of bison now support the ranching industry. Blue grama occurs statewide, and its flag-like flowers and seed heads are sometimes called "Grama's eyelashes."

Several butterflies depend on blue grama as the principal food for their caterpillars. Three examples are SIMIUS ROADSIDE-SKIPPER [*Notamblyscirtes simius*], UNCAS SKIPPER [*Hesperia uncas*], and RHESUS SKIPPER [*Polites rhesus*], all of which live on the shortgrass prairies. Go to grassy hilltops, such as Caprock Park south of San Jon in Quay County or around the Plains of San Agustín west of Socorro, to look for Rhesus in May and Simius in July. Colonies of Rhesus can be local and ephemeral, flying for as little as two weeks. Simius is widespread but never common. Watch for Uncas at nectar sources like thistles and milkweeds in May and again from July to September.

Blue grama also is part of broader caterpillar menus for Ridings' Satyr, Mead's Wood-Nymph, Garita Skipperling, Colorado Branded Skipper, Pahaska Skipper, Green Skipper, and others.

Top, left to right:
RHESUS SKIPPER
UNCAS SKIPPER
Left:
SIMIUS ROADSIDE-SKIPPER

Transition Zone

From top:
THICKET HAIRSTREAK
WESTERN PINE ELFIN

Journey to slightly higher country for the second tier of almost statewide butterflyscapes. Piñon pine, ponderosa pine, or Gambel oak woodlands and savannas cloak many of New Mexico's mid-level mountains and uplands. This vegetation marks the Transition Zone, which extends from roughly 6,000 feet to 8,000 feet in elevation. Piñon jays and Abert's squirrels bring these woodlands to life. If you think about it, this plant community thrives in much of the state except the eastern plains. I-40 traverses the Transition Zone from Grants to Gallup; I-25 penetrates its habitats from Santa Fe to Las Vegas. The towns of Raton, Ruidoso, Reserve, Red River, Mora, Angel Fire, Eagle Nest, Chama, Taos, Silver City, and Gallup are located in this life zone. Many of New Mexico's popular state parks and other recreation areas take advantage of the zone's pleasant summer weather. Butterflies that live here will not come to you; instead, you must seek them out.

Ponderosa pine [*Pinus ponderosa*] and piñon pine [*Pinus edulis*] serve as caterpillar food for three butterflies. Stroll through the open piney woods of Priest Canyon in the Manzano Mountains in middle-to-late May to see WESTERN PINE ELFINS [*Callophrys eryphon*] sipping nectar at understory flowers or perching on four-foot-tall pine saplings. The same flowers could have visiting THICKET HAIRSTREAKS [*Callophrys spinetorum*], which also sip

moisture from wet soil. While Pine Elfin caterpillars eat pine needles, Thicket Hairstreak larvae prefer dwarf mistletoes [*Arceuthobium*] that are parasites on ponderosa and piñon pines.

You have to come back in July or August to find PINE WHITE [*Neophasia menapia*], which is fairly easy to spot as it gathers nectar from streamside flowers such as cutleaf coneflower [*Rudbeckia laciniata*] or various asters. Larvae like ponderosa pine needles, but piñon is a second choice. The Lincoln National Forest around Ruidoso is an excellent locale for sighting Pine Whites.

This highly scented life zone has its contingent of butterflies that like open, dry habitats. The small, colorful FIELD CRESCENT [*Phyciodes pulchella*] frequents roadsides and fields within pine savannas. Look for adults nectaring unselectively at seasonal flowers at El Vado Lake State Park in May and June and again in August and September. The larvae eat asters.

Visit Madre Mountain in the Dátil Range in July to glimpse SQUARE-SPOTTED BLUE [*Euphilotes battoides*] and MEXICAN METALMARK [*Apodemia mejicanus*].

From top:

FIELD CRESCENT
PINE WHITE
MEXICAN METALMARK

Right:

SQUARE-SPOTTED BLUE

Gambel Oak

New Mexico State Parks

Top, left to right:
ROCKY MOUNTAIN DUSKYWING
SHORT-TAILED SKIPPER
Bottom, left to right:
ARIZONA SISTER
COLORADO HAIRSTREAK

Gambel oak [*Quercus gambelii*] is an important source of food for wildlife. Its acorns feed squirrels, birds, and bears; its leaves feed many moth and butterfly caterpillars.

April is the time to investigate Gambel oak groves for ROCKY MOUNTAIN DUSKYWING [*Erynnis telemachus*] and SHORT-TAILED SKIPPER [*Zestusa dorus*]. Adults of both species bask and perch on host oaks and come to moist soil to feed. Rocky Mountain Duskywings are fond of nectar, but Short-tailed Skippers prefer tree sap.

COLORADO HAIRSTREAK [*Hypaurotis crysalus*] flies in July and August. Adults bask in morning sunlight and then nap until evening, when they are most active. Males patrol clumsily about the oak canopy, often perching at head height. Adults visit moist earth.

ARIZONA SISTER [*Adelpha eulalia*] flies in June and August in the north but from May to October in the south. Males perch on high oak branches with their heads down, watching for females. They fly with several rapid wing beats followed by a distinctive, flat-winged glide before returning to their perches. Adults feed at wet sand.

From top:
WESTERN TAILED-BLUE
NORTHERN CLOUDYWING
ARACHNE CHECKERSPOT
RIDINGS' SATYR
AFRANIUS DUSKYWING

Both live in stands of antelope sage [*Eriogonum jamesii jamesii*]—the preferred caterpillar food—which in turn prefers disturbed, south-facing slopes in pine savannas. Adults never wander far from the host plant. Well-camouflaged RIDINGS' SATYR [*Neominois ridingsii*] is prevalent in grasslands and savannas. Flying weakly and jerkily near the ground, males search grassy knolls for females. Walk up to open hill summits to find ARACHNE CHECKERSPOT [*Poladryas arachne*]. Its larvae feed on beardtongues [*Penstemon*].

Legumes (that's the pea family) are an important plant group in most terrestrial ecosystems, and our ponderosa pine forests are no exception. Three species of butterflies feed herbaceous legumes to their young, fly in low areas, and frequent flowers and moist soil. NORTHERN CLOUDYWING [*Thorybes pylades*] flies in June in cool habitats; it has May and August flights farther south. AFRANIUS DUSKYWING [*Erynnis afranius*] flits about primarily in April and July. WESTERN TAILED-BLUE [*Cupido amyntula*] has a June flight in the north, with May and August flights in the southern mountains.

Grasses may be the most important plant group for butterflies in Transition Zone pine savannas. A healthy grassland understory is a sign that fire is playing its normal role in pine forests. While walking through such

[Continued on page 46]

Disturbance and Succession

For most butterfly species, their abundance and geographic extent boil down to how successful their larval host plants are. But plant communities are not static; they comprise populations of many species that confront countless, complex challenges of life. They continually grow, seek pollinators, disperse seeds, feed herbivores, die back in bad years, recover in good years, fight off diseases, reproduce, and die. Individual plants are immobile, but through their success or failure they influence other plants and animals in their vicinity. Interaction of all of these forces upon individuals and species causes all places to change over time. Slow or fast, apparent to us or not, plant communities are changing everywhere, always.

Succession is the gradual process by which a plant community changes over time following disturbance. Disturbances result from sudden catastrophes such as strong winds, floods, or fires. These events usually remove large, shade-making plants and make way for plants that thrive in open, sunny habitats. Growth of these pioneer plants is followed by establishment of longer-lived shrubs and then, in suitable areas, by trees. As they mature, trees shade out the smaller, sun-loving species that first colonized the site.

Succession in semi-arid New Mexico often takes decades or centuries for some forest communities. Though you might return to a particular place year after year, you could fail to recognize that seedlings have become trees and the area has transitioned from open and sunny to wooded and shady. The change in vegetation means a new menu of butterfly host plants and, consequently, a new suite of butterflies and other herbivores.

The next time you are in a native landscape try to look at it through your mental time machine. How might it have appeared 50 years ago? Do you see trees today that may have been absent then? Can you see seedlings today that may be full grown 100 years from now? Is there evidence of fire in the recent past? Imagine how that place changes over time in terms of plants and therefore in terms of butterflies.

Cycle of Disturbance and Plant Succession

Wildfire and Fire Ecology

Our chief source of information about wildfires is TV news, which tells a sensational, one-sided story that emphasizes negative, short-term consequences. Video footage shows flames consuming trees and threatening people's homes. Smoking stumps in a blackened landscape, however, are merely a prelude to a more interesting story.

Greater than collections of trees, forests are interdependent processes of plant succession, disturbance, and competition for resources. These processes operate and adjust, imperceptibly to us, over many decades. Wildfire is a natural disturbance that rejuvenates ecosystems by returning nutrients to the soil and restarting plant succession. It is essential and beneficial to New Mexico's grassland and ponderosa pine landscapes—and it is inevitable.

Many important caterpillar host plants respond quickly after a fire. Gambel oak resprouts vigorously within weeks by capitalizing on nutrient availability and abundant sunshine. New Mexico locusts, lupines, and other legumes thrive after fires. Perennial grasses also come back fast and green after a fire, as most ranchers know.

Butterflies have also adapted to wildfire. Several hairstreaks and blues survive by spending much of their time as pupae in the soil, where they are safe from fire's heat. Herbivores that eat fire-adapted and fire-dependent plants comprise a major part of the state's butterfly fauna.

When you see TV coverage of wildfires, remember that life will return quickly to burned areas. Plants and animals will re-occupy the land and thrive after a fire has ceased to be newsworthy.

New Mexico State Parks

Fenton Lake State Park, 2002

Above, from top:
MORRISON'S SKIPPER
PYTHON SKIPPER

[Continued from page 42]

areas, keep your eyes peeled for several resident butterflies whose larvae eat grasses.

Look for MORRISON'S SKIPPER [*Stinga morrisoni*] in spring: that means April in the south and May farther north. Males are persistent hilltoppers, so you can see them on Mount Sedgwick in the Zuni Mountains. Both sexes nectar at purple flowers like verbenas [*Glandularia*] and vetches [*Vicia*]. If you come upon some of these plants, stick around and watch for skipper visitors.

New Mexico's ponderosa savannas often become dry in June, so flower patches and mud holes are popular stops for butterflies that fly then. Take a jaunt to Water Canyon in the Magdalena Mountains. There you can expect ORANGE-HEADED ROADSIDE-SKIPPER [*Amblyscirtes phylace*] and PYTHON SKIPPER [*Atrytonopsis python*]. Both species look for mates by perching in gullies and low spots, and both are fond of nectar.

July is the time for CASSUS ROADSIDE-SKIPPER [*Amblyscirtes cassus*], which usually perches near the ground and feeds at nectar and water. Make a July

Left to right:
CASSUS ROADSIDE-SKIPPER
ORANGE-HEADED ROADSIDE-SKIPPER

Fendler's Buckbrush

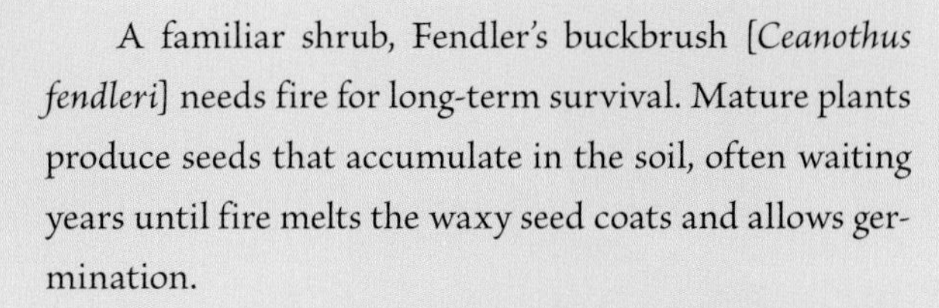

A familiar shrub, Fendler's buckbrush [*Ceanothus fendleri*] needs fire for long-term survival. Mature plants produce seeds that accumulate in the soil, often waiting years until fire melts the waxy seed coats and allows germination.

When amidst Fendler's buckbrush—on Burnt Mesa in Bandelier National Monument, for example—look for four butterflies whose larvae eat this plant. Two small species often perch on the host and nectar at its flowers. Green adults of BRAMBLE HAIRSTREAK [*Callophrys affinis*] fly in May and July. Perky NAIS METALMARK [*Apodemia nais*] adults flit about only in July. Walk to the top of the nearest rise, where males of PACUVIUS DUSKYWING [*Erynnis pacuvius*] appear in April and again from June through August. A fourth member of the buckbrush guild, CALIFORNIA TORTOISESHELL [*Nymphalis californica*], wanders far and wide. Watch for adults in June and July; a second generation flies in September, overwinters, and flies again in spring.

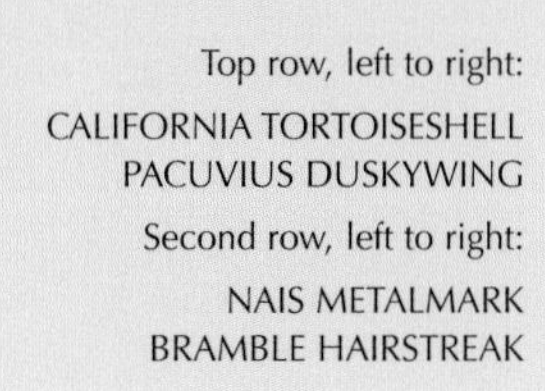

Top row, left to right:
CALIFORNIA TORTOISESHELL
PACUVIUS DUSKYWING
Second row, left to right:
NAIS METALMARK
BRAMBLE HAIRSTREAK

Above, from top:

CANYONLAND SATYR
NOKOMIS FRITILLARY

Right:

EDWARDS' SKIPPERLING

excursion to Three Gun Spring in the Sandía Mountains to see tiny EDWARDS' SKIPPERLING [*Oarisma edwardsii*] nectaring at flowers.

CANYONLAND SATYR [*Cyllopsis pertepida*] is on the wing in open pine savannas for much of the warm season, usually near watercourses with dappled shade. Adults fly in two or three successive generations and can be seen flopping in and out of bushes. This butterfly was a favorite object of study for novelist Vladimir Nabokov.

Streamside areas are habitats with specialized plants that in turn support a variety of herbivores, among which are several nifty butterflies. NOKOMIS FRITILLARY [*Speyeria nokomis*], perhaps New Mexico's most beautiful butterfly, is also one of its rarest, preferring wet meadows with host violets. Look for it along Negrito Creek east of Reserve in September.

Francis Snow

HISTORY HIGHLIGHT

Francis Huntington Snow led six entomological collecting expeditions to New Mexico between 1880 and 1894. Riding the new railroads to the western frontier, the University of Kansas entomologist sponsored these summer adventures to places where he and his students could discover and study new organisms.

Snow and his students went wherever the transcontinental railroad took them. In Santa Fe they explored the Santa Fe River Canyon in 1880. From Socorro they rode wagons to the Magdalena Mountains in 1881 and 1894. Above Las Vegas in 1882 and 1883, Snow found that Gallinas Canyon produced his "favorite objects of study in delightful variety and perfection." From Silver City he journeyed to the nearby Pinos Altos Mountains in 1884. These various excursions spanned elevations ranging from 4,500 feet to 10,000 feet.

Snow kept members of his expeditions busy all day and well into the night. Long after his young charges retired to their tents, he built roaring bonfires and collected moths attracted to the light. The explorers experienced flash floods and droughts, witnessed overgrazing by domestic livestock, and fled from Apaches defending their lands. Snow encouraged his students to collect anything and everything to improve the university's holdings, but the expeditions are best known for their collections of butterflies, moths, and beetles.

Francis Snow, University of Kansas professor of entomology

Snow made huge contributions to the study of New Mexico butterflies. He was the first to investigate compact geographic areas using multiple researchers over long periods. He documented 90 butterfly species, including the Green Skipper, which was previously unknown to science, and the Salome Yellow, which has not been seen here since. Through proper preservation and labeling of specimens, Snow's collections remain the oldest New Mexico material of real breadth and depth.

A July trip to Gallinas Canyon west of Las Vegas will almost always reward you with TAXILES SKIPPER [*Poanes taxiles*]. It darts about streamside habitats, perches with wings half open, nectars at meadow flowers, and siphons moisture from wet sand on hot days. It shares this habitat with distinctively gold-faced DUN SKIPPER [*Euphyes vestris*].

TAXILES SKIPPER

SILVER-SPOTTED SKIPPER [*Epargyreus clarus*] is hard to miss. This beauty lives where it finds plenty of New Mexico locust [*Robinia neomexicana*] for its caterpillars to eat. Adults fly in June, and males like to perch on protruding branches in stream corridors. If you find a patch of beebalm [*Monarda menthifolia*], give it some extra attention. Blooming in July, beebalm's elaborate, slender-tubed, pink flowers are perfect for long skipper proboscides.

Other streamside denizens among ponderosa woodlands become most visible in spring and again in late summer. MYLITTA CRESCENT [*Phyciodes mylitta*] is most obvious from April through May and July through August; its larvae eat thistles. Look for SATYR COMMA [*Polygonia satyrus*] adults in June, July, and August. A second generation in September and October hibernates and then flies again from March through May. Chief caterpillar foods are stinging nettle [*Urtica*] and hops [*Humulus lupulus*].

Above, left to right:

SATYR COMMA
DUN SKIPPER
MYLITTA CRESCENT

Below:

SILVER-SPOTTED SKIPPER

From top:

NORTHERN CRESCENT
DREAMY DUSKYWING

The third excursion into New Mexico's statewide butterfly landscapes climbs higher into the mountains, where you can encounter mixed conifer forests and subalpine meadows. These aptly named Canadian Zone habitats, ranging from roughly 8,000 feet to 10,000 feet, cover less territory than juniper-studded Upper Sonoran Zone grasslands and Transition Zone pine/oak savannas but are more widespread than you might guess. The Canadian Zone includes the expansive Jémez and Sangre de Cristo high country in north-central New Mexico. The Sacramento, Sierra Blanca, and Capitán ranges in the southeast push into this altitude bracket, as does the Gila high country in the west. Even some smaller, isolated ranges in the central part of the state penetrate this zone. Canadian Zone landscapes divide conveniently into four different butterfly habitat types: woodlands, meadows, stream corridors, and hilltops.

First explore the woodland component of this life zone in New Mexico. Woodlands can be shady, cool places that most sensible butterflies would avoid, so focus your efforts where the sun shines in forest openings or along forest edges. Aspen [*Populus tremuloides*] is a typical tree at this elevation, and it hosts DREAMY DUSKYWING [*Erynnis icelus*]. Flying in May and June, adults come to flowers and wet sand. In June and July, forest openings attract NORTHERN CRESCENTS [*Phyciodes cocyta*] seeking meadow flowers. Larvae eat plants in the daisy family, such as smooth aster [*Aster laevis*], and then overwinter.

HOARY COMMA

Open coniferous woodlands, such as those along Holy Ghost Canyon in the Sangre de Cristo Mountains above Pecos, are havens for HOARY COMMA [*Polygonia gracilis*]. Its caterpillars find good eating among currants and gooseberries. The summer generation appears from mid-June to mid-July. Offspring fly in August, overwinter, and fly again from mid-March to mid-May. Adults patrol creeks and feed at flowers, tree sap, and moist earth.

Sunny subalpine meadows are home to several summer butterflies. On your July morning stroll through knee-high grasses, perhaps at Eagle Nest Lake State Park, you might happen upon two skippers. Tiny GARITA SKIPPERLING [*Oarisma garita*] flies in July and often nectars at flowers. It is widespread in this habitat because its caterpillars eat a broad variety of grasses. TAWNY-EDGED SKIPPER [*Polites themistocles*] is larger than Garita and easy to distinguish. Adults frequent nectar and wet sand. The dark, quarter-sized critter flopping elusively among the grasses is probably the SMALL WOOD-NYMPH [*Cercyonis oetus*].

Among the grasses of most subalpine meadows, such as those near the summits of Mount Taylor, is a smattering of low cinquefoils [*Potentilla*]. The presence of these five-leafed plants offers a cue to look for the diminutive MOUNTAIN CHECKERED-SKIPPER [*Pyrgus xanthus*]. Adults are on the wing in May. They fly near the ground and, in the absence of nectar, are most visible at puddle edges.

From top:

MOUNTAIN CHECKERED-SKIPPER
GARITA SKIPPERLING
SMALL WOOD-NYMPH

Canadian Zone meadows often have a strong legume contingent, of which lupine [*Lupinus*] may be the most obvious. Lupine and other legumes host two of our finer blues, each of which feeds at flowers or sips moist soil. SILVERY BLUE [*Glaucopsyche lygdamus*] flies in May in the Gila country and Sacramento Mountains and then in June farther north. BOISDUVAL'S BLUE [*Plebejus icarioides*] adults are about in June and July.

This life zone receives more precipitation than lower elevations. Most places where you can park your car will be within a reasonable walk to a creek of some kind. If you visit Canadian Zone streams you should spot several interesting butterflies, since many deciduous trees and shrubs that serve as host plants thrive near watercourses.

From top:

SILVERY BLUE
BOISDUVAL'S BLUE
TAWNY-EDGED SKIPPER

Left to right:

WESTERN TIGER SWALLOWTAIL
SPRING AZURE

In springtime visit Bear Trap Campground in the San Mateo Mountains southwest of Socorro. April and May at this elevation usually reveal SPRING AZURE [*Celastrina ladon*]—a familiar flash of periwinkle that may be the first butterfly of the year. Come back here in June and July to find WESTERN TIGER SWALLOWTAIL [*Papilio rutulus*] wafting up and down stream corridors to seek mates and nectar. Moist earth at Bear Trap Springs attracts both species.

Streams in lower Canadian Zone areas produce a must-see summer butterfly scene. Spring Canyon in the Sacramento Mountains near Weed is a perfect spot. Look for stands of cutleaf coneflower near streams, seeps, or springs. Their bright yellow, daisy-like flowers bloom in July and can save you a lot of walking because they attract most butterflies within a quarter-mile radius. All you have to do is amble casually among the four-foot-high clumps and watch for the creatures, which should include WEIDEMEYER'S ADMIRAL [*Limenitis weidemeyerii*]. Males patrol stream corridors for mates, often returning to the same perch. Also look for SILVERY CHECKERSPOTS [*Chlosyne nycteis*], whose caterpillars eat coneflower foliage and whose adults nectar at the flowers. Sneezeweed [*Helenium*], another yellow, daisy-like flower, is just as popular among butterflies, but it grows at the upper end of the Canadian Zone.

From top:
SILVERY CHECKERSPOT
WEIDEMEYER'S ADMIRAL

Watch for butterflies siphoning electrolyte-rich fluids from stream edges on hot, sunny afternoons. These electrolyte parties may include MARGINED WHITE [*Pieris marginalis*], which flies near the ground in April, May, and June and again in July and August. Females place eggs on streamside mustards. MILBERT'S TORTOISESHELL [*Aglais milberti*] may elbow for room at the wet sand. Milbert's capitalizes on streamside habitats that grow its host, stinging nettle. Look for adults flying actively in June, July, and August; autumn flyers overwinter as adults and then fly again in March, April, and May.

From top:

MARGINED WHITE
MILBERT'S TORTOISESHELL

Left:

PERSIUS DUSKYWING

Do not leave the Canadian Zone without walking up to a prominent hill summit. Take a July jaunt to Sandía Crest near Albuquerque, when you can see each of the following three species whose males seek mates on hilltops. PERSIUS DUSKYWING [*Erynnis persius*] females place eggs on legumes. WEST COAST LADY [*Vanessa annabella*] is less common than the other "ladies," preferring higher mountains. MEXICAN CLOUDYWING [*Thorybes mexicana*] adults fly mainly in June and July. Larvae eat legumes. All three of these butterflies visit flowers.

From top:
WEST COAST LADY
MEXICAN CLOUDYWING

Finding a Mate

If you could ask them, males of any species would confirm that mate availability is a critical element of habitat. Mating is the primary function of adult butterflies; it is the only reason for them to be nourished, hydrated, electrolyte-balanced, warm, and dry. But the universe is large and butterflies are small, so the challenge of finding a suitable mate is great—like finding a needle in a haystack. Humans narrow down the universe by seeking mates, for example, at workplaces, schools, places of worship, or bars. Butterflies employ mate location strategies linked to natural landscape features such as host plants, hilltops, or gully bottoms.

Adults of some species seek mates in close proximity to larval food plants. Males perch on prominent twigs or branches and occasionally fly in brief sorties around individual plants or groups of plants. Host plants also provide food and shelter. These butterflies find mates by staying near home and seeking the girl next door.

Adults of other butterfly species are hilltoppers: they narrow down the universe by looking for mates on topographic high spots. Males employing this approach typically fly uphill until they reach the top. They perch where they have a good view and keep a sharp lookout, flying out to investigate anything that comes near. If it is not a female of their species, they chase away the intruder. They are belligerent toward males of the same species, butterflies of other species, other insects, birds, and even nosy butterflyers. If they see a female of their species, they court and attempt to mate. Due to so much mate-locating activity, hilltops often reward diligent climbers with butterflies not observable at locales that require less exertion.

Another mate location strategy involves adults of certain species that fly downhill until they reach the bottom of the landscape. Usually this is a swale, gully, valley, or riverbed. Males then perch on streamside boulders or overhanging branches to get the best view of the stream corridor. From there they fly upstream and downstream to find mates—a behavior called patrolling. As on a hilltop, a male darts out after passersby, chasing away competing males in spiraling dogfights. The vanquished move on, while the victor returns to the favored perch and resumes its quest.

This book, so far, has offered you the equivalent of four ingredients to a tantalizing butterfly stew. The first, most basic ingredient consisted of those species that are easy to see, common, widespread, and statewide—the broth. Then, based on elevations and life zones, you got a taste of three groups and roughly 60 butterflies that are routine only in their preferred Upper Sonoran, Transition, or Canadian life zones—the potatoes, onions, and carrots.

Next comes the spicy chile to flavor the stew. Chile is different wherever you go in New Mexico, and so are the butterflies. An observant journey around the state reveals distinctive ecosystems in various areas. Each regional ecosystem is unique in terms of climate, landforms, plants, and animals—including butterflies.

Now prepare to embark on a counterclockwise journey through New Mexico beginning in the eastern plains. For each geographic region (see map at right), the discussion will highlight butterflies that make the area special and worth visiting. In some sections you'll find an in-depth examination of one particular ecoregion, as in the first stop in the eastern plains, which is part of the larger Great Plains ecoregion. Or you may discover that the inextricable intertwining or convergence of larger continent-scale ecoregions makes some other part of the state quite distinct. All the while, keep in mind that the regional butterfly species build on the array of ubiquitous ones found statewide as well as the life-zone-specific species described in previous sections.

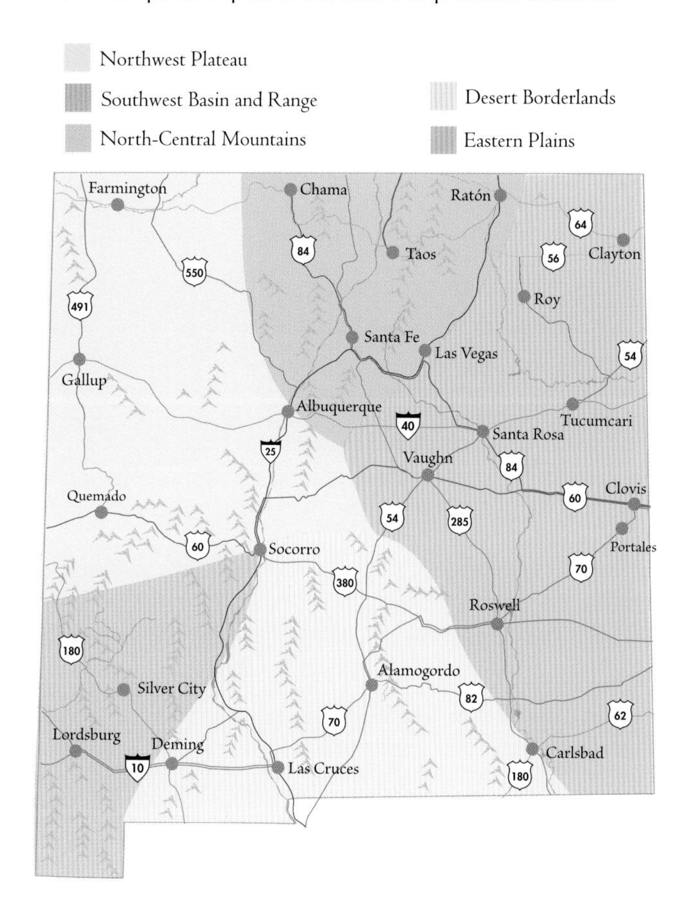

MAP OF ECOREGIONS

- Great Plains
- Rocky Mountains
- Colorado Plateau
- Mogollón Highlands
- Sierra Madre Uplands
- Chihuahuan Plateau

New Mexico's Ecoregions

Ecologists and geographers divide North America into distinct areas called ecoregions. Each ecoregion has characteristic rocks, landforms, and climate conditions that foster a distinct assemblage of plants and animals. New Mexico lies at the juncture of several ecoregions.

The Great Plains ecoregion typifies North America's heartland and reaches into New Mexico's eastern plains. The eastern grasslands are shortgrass versions of prairies that graduate to tall grasses east of the Mississippi River. The gentle terrain transitions westward into mesas and valleys.

The Colorado Plateau ecoregion, part of the Great Basin of Nevada and Utah, occupies northwest New Mexico, where the San Juan River drains toward the Colorado River. The low, dry terrain is characterized by arid shrublands within a landscape of mesas and valleys bordered by mountains and high plateaus.

The Chihuahuan Plateau ecoregion extends from northern Mexico into southern New Mexico. It includes internally drained basins, river valleys, and low mountains. Though often dismissed as mere desert, the grasslands, shrublands, and river corridors support great biological diversity.

The Rocky Mountains ecoregion extends from Alaska south along the spine of western North America to New Mexico. The state's highest and coldest ecoregion, it dominates northern New Mexico with conifer forests and alpine meadows that reach south to the Sacramento Mountains.

The unique ecological influence of northern Mexico's Sierra Madre Uplands ecoregion extends into southwest New Mexico's Bootheel. Though occupying only a small part of our state, it makes a proportionally large contribution to New Mexico's butterfly fauna.

The rugged country of the Mogollón Highlands ecoregion separates the Colorado Plateau and Chihuahuan Plateau, while loosely connecting the southern Rocky Mountains and Sierra Madre Uplands. This area of ecological mixing is known as Gila country.

Places where ecoregions converge are diverse areas for butterflies because species of each ecoregion fly together. The United States has five areas of major butterfly diversity. New Mexico is fortunate to share three of these biodiversity hot spots with neighboring states.

For more information about biodiversity hot spots see the listing under Biodiversity in the Recommended Reading section on page 149.

REGIONAL SPECIALTIES

Eastern Plains

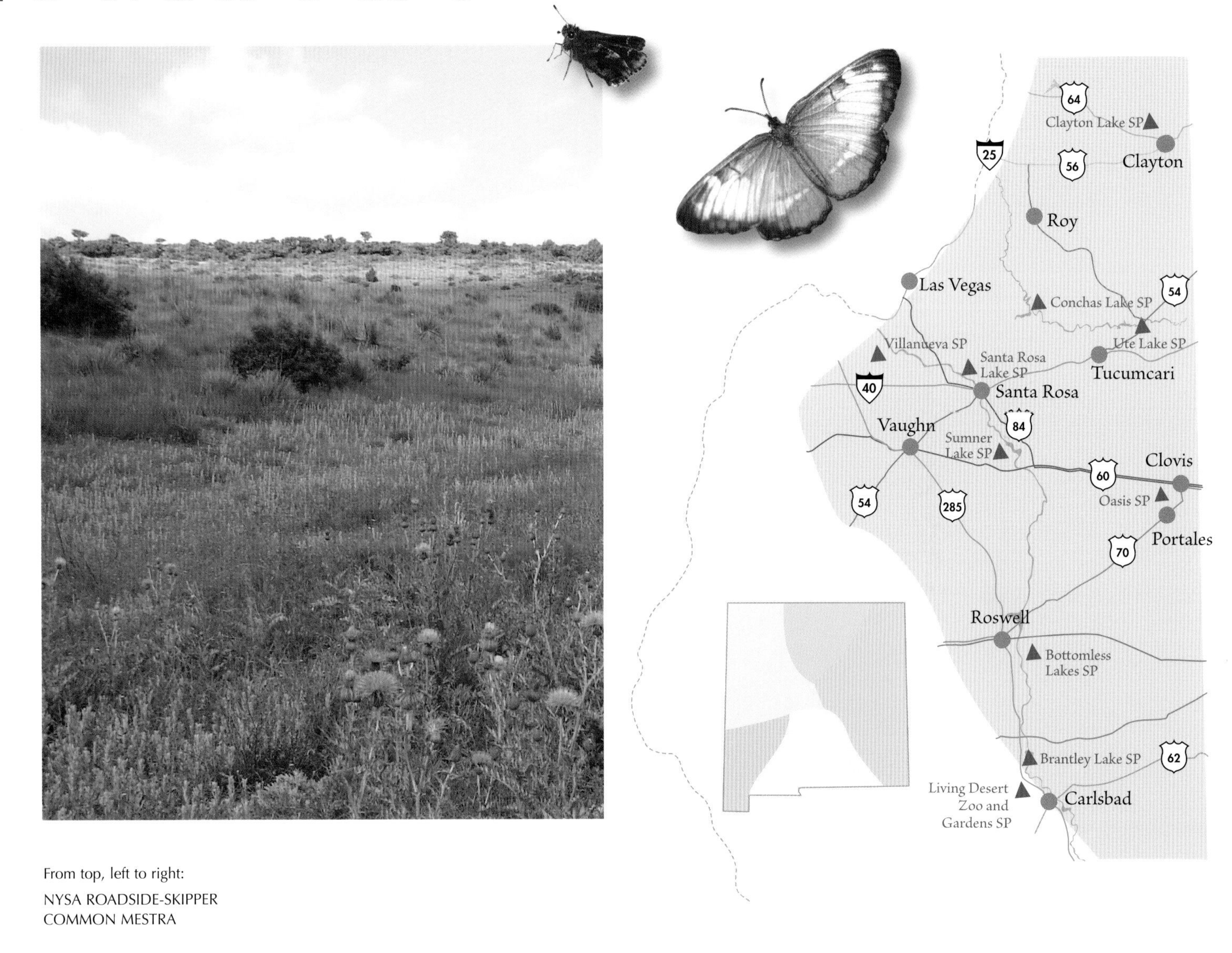

From top, left to right:

NYSA ROADSIDE-SKIPPER
COMMON MESTRA

Butterfly species typical of the Great Plains ecoregion dominate New Mexico's eastern section. This largely treeless area is home to many grasses and other herbaceous plants. Though the eastern plains may appear somewhat uniform throughout, a keen eye reveals great variation from south to north. Species diversity mirrors topographic diversity, and this area has less of both than anywhere else in New Mexico. Nevertheless, the region's ridges, streamsides, and escarpments are worth exploring for butterflies. The discussion below features several specific locations to visit, starting in the south and moving northward through the plains.

A trek in search of plains butterflies could include a May trip to Bitter Lake National Wildlife Refuge near Roswell. Ungrazed grasslands there and at Bottomless Lakes State Park offer excellent habitat for southern skippers. Along sandy jeep trails or arroyos that feed the main valley, you can find male NYSA ROADSIDE-SKIPPERS [*Amblyscirtes nysa*] perching and patrolling for females. In late summer the southern prairies sometimes are decorated with a subtropical visitor—COMMON MESTRA [*Mestra amymone*]. Look for it in moist oases, flying weakly and siphoning nectar from flowers.

Continue up the Pecos River Valley to Sumner Lake State Park, a fertile locale throughout the warm season. The best overall habitat is in the canyon below the dam, where native plant communities remain healthy. PEARL CRESCENT [*Phyciodes tharos*] prefers these riverside areas. Look around the lake margins for PHAON CRESCENT [*Phyciodes phaon*]. Larvae eat mat-plant [*Phyla lanceolata*], which likes wet soil. Phaon adults should be nearby, flying near the ground.

Above:
PHAON CRESCENT
Right:
PEARL CRESCENT

Along the state's eastern border, the landscape smoothes out and drainages dip gently toward Texas. Cultivated agriculture becomes more prevalent and public land scarcer. Ned Houk Park in Clovis offers access to undeveloped lands where you can see butterflies typical of the eastern United States, like QUESTION MARK [*Polygonia interrogationis*], which seems most widespread in September. The species became more common here after Governor Clyde Tingley gifted New Mexico's treeless, Dust Bowl-afflicted eastern plains with 25,000 Siberian elms in the 1930s. Question Mark caterpillars consider this plant a delicacy. LITTLE YELLOWS [*Pyrisitia lisa*] enliven late summer fields across the agricultural region. This eastern U.S. butterfly shows up most frequently in September and October.

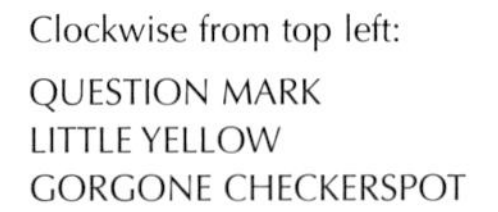

Clockwise from top left:

QUESTION MARK
LITTLE YELLOW
GORGONE CHECKERSPOT

Also check out Oasis State Park near Portales, where GORGONE CHECKERSPOTS [*Chlosyne gorgone*] rule in September. Gregarious Gorgone larvae gobble blossoms of prairie sunflower [*Helianthus petiolaris*], Gorgone chrysalids cling to stucco walls, and Gorgone males quarrel over perching rights at the top of a prominent sand dune.

Two other butterflies are common in this region in July. Golden DELAWARE SKIPPERS [*Anatrytone logan*] fly in June and July. Though they favor wet areas and feed associated grasses to their larvae, adults spread out widely. Any thistle patch in the eastern plains is likely to have a few Delaware Skippers by mid-July. Another regular is the day-glo-orange GULF FRITILLARY [*Agraulis vanillae*]. Though it does not breed or overwinter here, it often flies into this area. Adults frequent riparian habitats to find food, water, and shelter.

From top:

DELAWARE SKIPPER
GULF FRITILLARY

Venturing north from the irrigated Clovis-Portales area, you eventually reach the northern edge of the Llano Estacado and encounter eastern New Mexico's most daunting physiographic feature—the Caprock Escarpment. Not scared? Good, because the escarpment offers some of the best butterfly habitat in eastern New Mexico. Its north-facing slopes are hospitable to plants and animals that normally live at higher, cooler sites; hence, butterfly diversity is great.

Here you will discover OLYMPIA MARBLE [*Euchloe olympia*] flying in April and May, often near ridges or mesa tops. Larvae eat mustards. Come back in August to find HORACE'S DUSKYWING [*Erynnis horatius*], which lives on wavy-leaf oaks [*Quercus undulata*] that flourish on the escarpment. Adults perch on ridgetops or nectar at roadside flowers. You can access these habitats at Caprock Park, south of San Jon in Quay County. Or you can study maps to locate other public lands on the escarpment.

From top:

OLYMPIA MARBLE

HORACE'S DUSKYWING

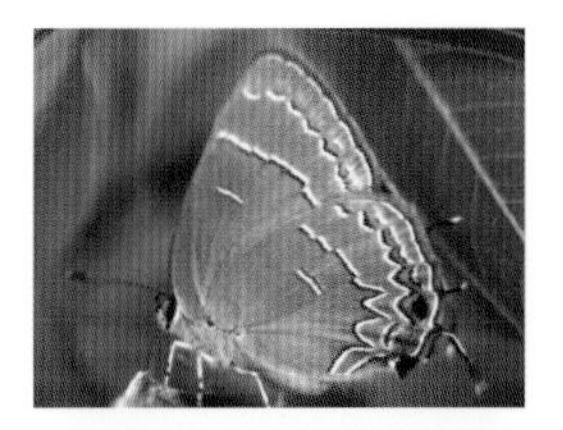

North of the Caprock Escarpment you enter the mesa-valley terrain of northeast New Mexico. The prominent valley was forged by the Canadian River, which begins on the east flank of the Sangre de Cristo Mountains and drains east toward the Mississippi River.

Within this two-tiered landscape, Conchas Lake and Ute Lake state parks near Tucumcari provide access to lowland and riparian habitats. As with most reservoirs, the best habitats occur below the dam, where the natural terrain remains rough and uninundated. Here you will find soapberry trees [*Sapindus drummondii*], which host SOAPBERRY HAIRSTREAK [*Phaeostrymon alcestis*]. Flying in June, adults perch on the small trees and seek nearby nectar. White flower clusters atop soapberries also attract other butterflies.

Away from the main valleys are widespread flats covered with sprawling stands of soapweed yucca [*Yucca glauca*], where you might spot STRECKER'S GIANT-SKIPPER [*Megathymus streckeri*]. Active in May, adults perch on the ground or on low shrubs and fly in agitated circles when disturbed.

From top:
SOAPBERRY HAIRSTREAK
STRECKER'S GIANT-SKIPPER

From top:
HACKBERRY EMPEROR
DOTTED CHECKERSPOT
SOUTHERN HAIRSTREAK

Zigzag across eastern New Mexico to a more upstream location on the Canadian River. Mills Canyon, within the Kiowa National Grassland, is managed by Cibola National Forest. Just northwest of Roy, the canyon offers a perfect locale to explore beautiful landscapes and butterflies distinctive to this part of the state.

Start by examining ridgetops along the access road for DOTTED CHECKERSPOT [*Poladryas minuta*], which lives here by virtue of its fondness for various prairie beardtongues [*Penstemon*]. Adults appear in May and again in August and September. Look for them at nectar and on hilltops.

The dirt road from mesa top to valley bottom can be challenging, so check conditions before setting out. This rugged road traverses some of the best habitats at Mills Canyon. Walk along it and explore associated creeks, draws, and nectar patches for some special butterflies. Among these is SOUTHERN HAIRSTREAK [*Satyrium favonius*], which occupies brushy, oak-filled canyons and hillsides. Adults fly in June and love dogbane nectar [*Apocynum*]. You'll see netleaf hackberry [*Celtis reticulata*] alongside abundant wavy-leaf oak in the canyons, so keep a sharp eye out for HACKBERRY EMPEROR [*Asterocampa celtis*]. This critter flies primarily in May and June and again in September. Males patrol gullies and perch head-down on canyon walls and trees, but females stay near hackberry trees.

Different habitats greet you on the broader valley bottom. Here COMMON WOOD-NYMPH [*Cercyonis pegala*] is a familiar sight in moist grasslands. Adults fly in July and August, lurching awkwardly through grasses and shrubs but stopping to siphon plant sap and nectar. Summer brood GOATWEED LEAFWING [*Anaea andria*] also should be here. Larvae eat doveweed [*Croton texensis*], which grows in local low spots. You can glimpse this butterfly again in September and October or March through May after it emerges from hibernation. Adults eat tree sap and perch like dead leaves in deciduous trees. Finally, VICEROY [*Limenitis archippus*] is a rare find near streamside willows in summer.

From top:

COMMON WOOD-NYMPH
GOATWEED LEAFWING

Left:

VICEROY

Clayton Lake State Park in far northeast New Mexico is a must-see destination. Because of its northeastern location, the park yields butterflies rarely found elsewhere in the state. A late May visit will net (no pun intended) LEAST SKIPPER [*Ancyloxypha numitor*], normally a species of the eastern United States. It likes habitats like those along Seneca Creek below Clayton Lake. Look for adults perching, nectaring, and dodging among reeds during July and August.

Lake margin habitats at Clayton Lake have sufficient types and numbers of docks [*Rumex*] to satisfy the needs of two coppers that find few other homes in New Mexico. A Great Plains butterfly, GRAY COPPER [*Lycaena dione*], occurs along the southwest margin of the lake. Watch for it in July. BRONZE COPPER [*Lycaena hyllus*] is a regular marsh and swamp dweller in the northeast United States, but peripheral western colonies dot the prairies east of the Rockies and south as far as northeast New Mexico. Adults fly in July and August, darting through wet meadows, perching in tall grasses, and sipping nectar. This species also occurs at Maxwell National Wildlife Refuge south of Raton.

From Clayton you can't go much farther east or north, so head west. A short drive takes you into a region that fits as well with the Rocky Mountains as it does with the Great Plains. Stretching from Raton to Clayton, the volcanic

Clockwise from above:

GRAY COPPER
BRONZE COPPER
LEAST SKIPPER

From top:
ALBERTA ARCTIC
HOBOMOK SKIPPER

landscapes of northeast New Mexico offer unusual cinder cones and basalt-capped mesas. Mesa side slopes are wooded like the Rocky Mountains, but mesa tops support cool grasslands typical of Canadian prairies. This rugged volcanic landscape harbors eastern U.S. butterflies stranded hundreds of miles or more from their sisters.

Sugarite Canyon State Park near Raton is an ideal place to investigate these special habitats, which straddle two regions—the eastern plains and north-central mountains. Late May is cool in northeastern New Mexico but a good time to drive up Soda Pocket Road and park at the trailhead for the Ponderosa Ridge and Opportunity trails. The left-hand trail wanders slowly uphill along Soda Pocket Creek among pines and thick undergrowth of Gambel oak and chokecherry. Along the way you might catch sight of HOBOMOK SKIPPER [*Poanes hobomok*], which is predominantly found in the eastern United States. Adults nectar at flowers and sip moisture from creek banks.

When you come to the fork in the trail, bear right toward Little Horse Mesa. If you can, clamber to the top, walk out onto the mesa, and enter another world—a high, flat grassland. From here you can see grassy mesa tops stretching from Fishers Peak near Trinidad, Colorado, east toward Clayton. These prairies are home to ALBERTA ARCTIC [*Oeneis alberta*], which likes the windswept subalpine grasslands. Larvae eat fescue grasses, and adults fly among bunchgrasses in May after snow disappears. Adults do not nectar, so kick your way through the grasses to make them fly.

While walking the Sugarite trails in July you might spot a butterfly that lives among these mesas and nowhere else. 'RATON MESA' NORTHWESTERN FRITILLARY [*Speyeria hesperis ratonensis*] is our palest race of Northwesterns. Its caterpillars eat Canada violets [*Viola canadensis*] in the pine-oak understory. A grass feeder, CROSSLINE SKIPPER [*Polites origenes*], flies from late June to July in grassy swales. Like Hobomok, this eastern U.S. butterfly has satellite colonies along the Rocky Mountain Front Range from North Dakota to New Mexico. One last rare treat here is MOTTLED DUSKYWING [*Erynnis martialis*]. Mottled repeats the geographic pattern of Hobomok and Crossline. Larvae eat Fendler's buckbrush. Look for adults on hilltops in June and September.

Hobomok, Crossline, and Mottled lived farther south and east on the plains during the last Ice Age, about 15,000 years ago. Warming and drying climates starting 10,000 years ago forced these creatures uphill and northward to remain in conditions they (and their host plants) liked. Now the descendants of these prehistoric butterflies find themselves at elevations of 8,000 feet to 9,000 feet in Colfax County, while the plains their ancestors once occupied are warmer, drier, and home to other butterfly species. This phenomenon may offer a clue to what lies ahead for our butterflies in a warming world.

From top:
'RATON MESA' NORTHWESTERN FRITILLARY
CROSSLINE SKIPPER
Opposite page, top:
MOTTLED DUSKYWING

Benjamin Hyde and F. Martin Brown

HISTORY HIGHLIGHT

Benjamin Talbot Babbitt Hyde grew up in New York City as a Harvard-educated heir to the Babbitt Soap Company (remember Bab-O soap?). Curious, worldly, and philanthropically inclined, Hyde sponsored archaeological expeditions to Chaco Canyon in the 1890s, collected natural history specimens in the Four Corners region, and became an experienced outdoorsman.

Benjamin Hyde and his wife, Helen, in Santa Fe circa 1930

Upon his return to New York, Hyde was dismayed to find city kids afraid of nature and the outdoors. By providing hands-on learning opportunities—often pulling live snakes out of deep coat pockets—he became the prototypical park naturalist. Hyde lobbied to make nature study a key part of outdoor experiences for New York youths at Bear Mountain–Harriman Park and convinced early Boy Scout leaders to incorporate such learning into their educational mission. He returned to New Mexico in 1927 and established the Children's Nature Foundation in Santa Fe. As "Uncle Bennie," he led youngsters into the mountains to learn about the outdoors. Upon Hyde's death in 1933, his widow donated his 300-acre refuge above Santa Fe to become Hyde Memorial State Park.

Hyde's passion for teaching children about nature rubbed off on F. Martin Brown, who in his youth worked as a naturalist at Harriman Park and developed insect research programs in the early 1920s. Brown eventually moved west, where he had a long and distinguished career as a science teacher at Fountain Valley School in Colorado Springs. "Brownie" also was a lepidopterist who published numerous scientific and popular articles about butterflies, co-authored the first book on Colorado butterflies in 1957, and became the honorary dean of Rocky Mountain butterflies. Among his exploits was the discovery of the Capulin Mountain race of Alberta Arctic.

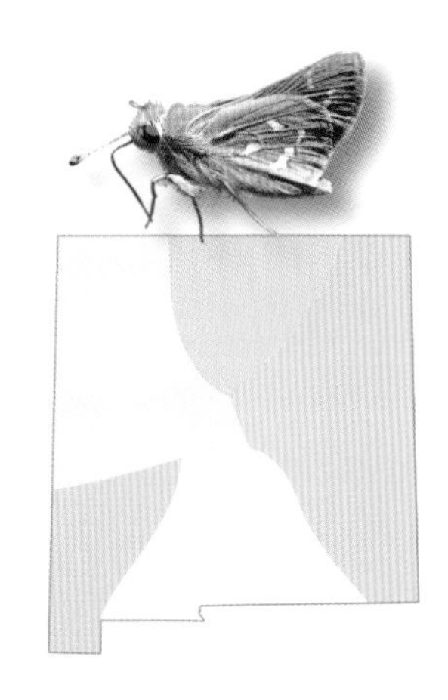

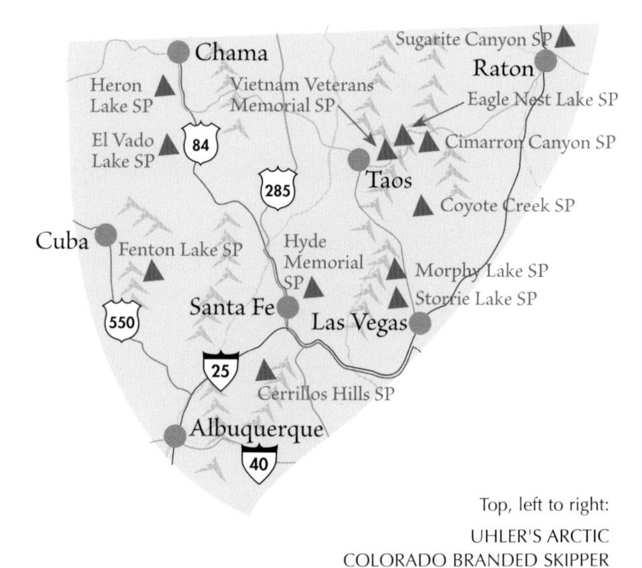

Top, left to right:

UHLER'S ARCTIC
COLORADO BRANDED SKIPPER

The Rocky Mountains ecoregion in north-central New Mexico encompasses the state's most mountainous landscape. Elevations range from 6,500 feet to 13,150 feet. Boasting the Land of Enchantment's highest terrain and coldest climates, the area contains Transition, Canadian, Hudsonian, and Arctic life zones. This southern extension of the Rocky Mountains separates the Great Plains and Colorado Plateau ecoregions and supports one of the richest communities of butterflies in New Mexico.

Within this region, you can begin your journey in the east-west uplands of the Raton-Clayton volcanic field. Jutting into the plains, these mesas have north-facing slopes where colder life zones and habitats can persist despite the latitude and lower altitude. The area harbors not just remnant populations of eastern U.S. butterflies but also fringe populations of colder Rocky Mountain species, making it a fascinating destination for butterfly lovers.

Sugarite Canyon State Park, the last plains site explored in the previous section, is also your first stop in this mountainous region. The high, grassy mesas and meadows have varying slopes, moisture, and grasses. Differences in the grasslands that seem subtle to you are quite apparent to butterflies found in this area. COMMON RINGLET [*Coenonympha tullia*] flies in June and July where grasses mix with trees. UHLER'S ARCTIC [*Oeneis uhleri*] adults appear in June. They perch in grass clumps and fly low over grassy high spots. The species is not common, but you may get lucky if you look atop Little Horse Mesa and higher mesas in the area. These same mesas are alive with COLORADO BRANDED SKIPPER [*Hesperia colorado*] in August. Your last stop at Sugarite is along Chicorica Creek behind the park's visitor center, where MEXICAN SOOTYWING [*Pholisora mejicana*] flies near the ground to sip nectar and moisture. Numbers peak in May and July. Mexican Sooties share the same pigweed [*Chenopodiaceae*] hosts as Common Sooties, and the two occur together here.

Top, left to right:
COMMON RINGLET
MEXICAN SOOTYWING

From top:
EDWARDS' FRITILLARY
APHRODITE FRITILLARY
'FRONT RANGE' NORTHWESTERN FRITILLARY
GREEN COMMA

U.S. 64 traverses scenic Cimarron Canyon between Cimarron and Eagle Nest. Lands straddling the canyon are managed by the New Mexico Department of Game and Fish as the Colin Neblett Wildlife Area, and New Mexico State Parks oversees recreation facilities along the valley bottom at Cimarron Canyon State Park. Violets [*Viola*] growing on adjacent forest floors provide caterpillar food for fabulous fritillaries, three of which are visible in July. EDWARDS' FRITILLARY [*Speyeria edwardsii*] adults search for nectar and patrol hilltops; their flight style is fast, furious, and far ranging. APHRODITE FRITILLARY [*Speyeria aphrodite*] is a familiar sight in the eastern United States, but its larger distribution includes the southern Rocky Mountains. Aphrodite adults passionately seek nectar from cutleaf coneflower. 'FRONT RANGE' NORTHWESTERN FRITILLARY [*Speyeria hesperis electa*] adults feed at wet sand and milkweed [*Asclepias*] flowers.

Three additional butterflies abound in late summer. 'FRONT RANGE' ANICIA CHECKERSPOT [*Euphydryas anicia capella*] occurs from Las Vegas, New Mexico, north into Colorado. This reddish race flies in late July and August and is most evident at Coyote Creek State Park near the group shelter in the afternoon. Look for GREEN COMMA [*Polygonia faunus*] in riverside openings within mixed conifer woodlands. Two good spots are the Big Tesuque Campground along the Santa Fe Ski Basin Road or near Holy Ghost Campground north of Pecos. Adults emerge in late summer, fly in August and September, and then hibernate; survivors fly again and mate in May and June. PURPLISH COPPER [*Lycaena helloides*] prefers open streamside habitats. It is ecologically versatile and pervasive in north-central New Mexico, occupying a wide range of elevations. Adults flit about from June to August, avidly seeking nectar and water.

Butterfly landscapes distinctive to the Sangre de Cristo Mountains express themselves at high elevation, where treeline begins at 11,000 feet and the air gets a bit thin. Above that point you encounter arctic and subarctic habitats that are scarce in New Mexico. Trees disappear; rocks, shrubs, forbs, sedges, and lichens dominate the terrain. Tundra plants support their own assemblage of butterflies adapted to life in this cold, windy place. Life cycles for many butterflies require two years here because the growing season is so short.

Visit the base of the Santa Fe Ski Basin or similar locations in the Sangre de Cristos at elevations from 10,000 feet to 11,000 feet. If you investigate wet spots such as willow bogs, you might catch a glimpse of SCUDDER'S SULPHUR [*Colias scudderii*]. It flies in July and August, preferring wet meadows with scrubby willows [*Salix*]. Take an excursion to Hamilton Mesa

From top:
SCUDDER'S SULPHUR
PURPLISH COPPER
'FRONT RANGE' ANICIA CHECKERSPOT

HISTORY HIGHLIGHT

Theodore Cockerell

Entomologist Theodore Dru Alison Cockerell was born in England in 1866. Nagged by tuberculosis in his early 20s, he sought employment in the American Southwest for its curative climate. Cockerell had a considerable background in field biology, including service under Alfred Russell Wallace, co-father (with Charles Darwin) of the modern theory of evolution. He angled for a post in New Mexico and arranged to take Charles Townsend's place at the College of Agricultural and Mechanical Arts (now New Mexico State University) in 1893. As professor of entomology and zoology, Cockerell studied agricultural pests and taught entomology. He documented the life history of the Western Pygmy-Blue and the color forms of the Bordered Patch.

In 1899 Cockerell transferred to the Normal School in Las Vegas. He and his wife, biologist Wilmatte Porter, hosted the nation's leading entomologists at their home in Sapello in 1900 and 1901. Their collective effort led to the discovery of new forms of Nokomis Fritillary and Scudder's Sulphur.

Theodore Cockerell as a young man

Cockerell had a positive impact on entomology in New Mexico, but his views on higher education clashed with school politics. In 1903 he led a revolt that ended in mass resignations among the school's faculty, president, and regents. In 1904 Cockerell moved to Colorado, where he had a long, prominent career studying bees. The Normal School re-opened in 1915 as Highlands University.

or Spring Mountain in the Santa Fe National Forest of western San Miguel County in late July or early August to replicate Theodore Cockerell's first state sightings of several alpine butterflies. In these high, flowery meadows he caught MORMON FRITILLARY [*Speyeria mormonia*], ARCTIC FRITILLARY [*Boloria chariclea*], and GRIZZLED SKIPPER [*Pyrgus centaureae*] in early August 1900.

Throughout this treeline landscape and higher, SMINTHEUS PARNASSIAN [*Parnassius smintheus*] dominates the scenery. Look for adults gathering nectar from alpine flowers in July. Males wander patiently in search of females. Larvae eat succulent stonecrops [Crassulaceae]. Large, white, mobile, and visible, Smintheus is the most prominent species of this rarified butterflyscape.

WESTERN WHITES [*Pontia occidentalis*] occupy an odd span of habitats, from Transition Zone high plains to Hudsonian Zone treeline. Larvae eat mustards and more. Adults hilltop on Wagon Mound in spring and on Gold Hill near Wheeler Peak in summer.

Treeless summits of New Mexico's highest mountains seem unlikely places for butterflies, but several specialized species thrive here. Most of these habitats are wilderness areas accessible only to climbers with strong legs and lungs—an exhilarating challenge! Plan a mid-July expedition to Wheeler Peak, New Mexico's highest summit. A steep five-mile walk each way awaits you, with 4,000 feet

Clockwise from top left:
GRIZZLED SKIPPER
MORMON FRITILLARY
SMINTHEUS PARNASSIAN
WESTERN WHITE
ARCTIC FRITILLARY

From top:

MAGDALENA ALPINE
LUSTROUS COPPER
MEAD'S SULPHUR

of elevation gain. If you start at dawn, you'll reach high tundra habitats by late morning and can start down before lightning makes life a bit too stimulating on the mountaintop.

As you emerge from the spruce and fir forest that dominates the first hour or two of your uphill walk, you'll notice that rockslides encrust Wheeler's slopes. Also called talus or scree, these rocks and boulders are fairly stable, as suggested by the rich surface growth of lichens. But they are still wobbly enough to warrant care about where you place your feet.

If you clamber carefully among the scree, you may catch sight of two rockslide denizens. Search among the boulders for MAGDALENA ALPINE [*Erebia magdalena*], the state's blackest butterfly. Its host grasses and sedges live in the crevices between the rocks. If you startle a Magdalena Alpine, don't chase it across the treacherous talus. You don't need a broken ankle just now! Also be aware of any small flashes of burnt orange, which may prove to be LUSTROUS COPPER [*Lycaena cupreus*]. This tiny creature likes to perch in grassy swales between the boulevards of boulders. New Mexico populations have been found only on the side slopes of Wheeler Peak.

MEAD'S SULPHUR [*Colias meadii*] thrives here, too; it occupies swales in high tundra ridges just above treeline. Adults fly in July, often coming to nectar. If you disturb one, watch it carefully as it flies to the ground and disappears among low foliage. Caterpillars eat clovers.

While the sun still shines, climb to Wheeler's summit to see three arctic butterflies. Check for marmots and forget-me-nots as you gasp, puff, and catch your breath. Adult POLIXENES ARCTIC [*Oeneis polixenes*], MOTTLED ARCTIC [*Oeneis melissa*], and 'ALPINE' ANICIA CHECKERSPOT [*Euphydryas anicia eurytion*] cruise back and forth along the most exposed, rocky ridges, searching for females. You also can spot some of these species on the Truchas peaks and Santa Fe Baldy.

West of the Río Grande, the Jémez and Tusas mountains are less rugged than the Sangre de Cristo Mountains. Gentler, rounded slopes encourage snowmelt and rainfall to percolate slowly into the soil and emerge later to feed wet meadows, springs, streams, and wetlands. These moist areas provide homes for butterflies that cannot find suitable habitats in the more vertical Sangre de Cristos.

From top:

POLIXENES ARCTIC
MOTTLED ARCTIC
'ALPINE' ANICIA CHECKERSPOT

To explore butterflies here, venture first into the Jémez Mountains. Mid-May is the best time to catch sight of HOARY ELFIN [*Callophrys polios*]. Resembling chips of pine bark, Hoaries cavort close to the ground, perching and nectaring near their sprawling host, kinnickinnic or bearberry [*Arctostaphylos uva-ursi*]. Look for them in open pine savannas near Fenton Lake State Park. On the same May trip you can visit Burnt Mesa at Bandelier National Monument, where plant succession since the 1977 fire provides excellent habitat for DUSTED SKIPPER [*Atrytonopsis hianna*]. This herbivore on little bluestem grass [*Andropogon scoparius*] is a species of the eastern U.S. prairies that has satellite populations in this region. Adults perch in gully bottoms and nectar at iris and thistles (all purple!).

Below, left to right:
DUSTED SKIPPER
HOARY ELFIN

Moving up in elevation, CHRYXUS ARCTIC [*Oeneis chryxus*] prefers bunchgrass meadows near treeline. Adults fly in June and early July. Camp May, west of Los Alamos and next to the Pajarito Mountain Ski Area, is an excellent place to watch these critters glide across south-facing alpine grasslands. See how well disguised they are when perched on the ground? Another butterfly to seek at this time is PALE SWALLOWTAIL [*Papilio eurymedon*], which stakes out hilltops and sips nectar or moisture. Fendler's buckbrush is the primary host, but wild cherries will do in a pinch.

Also representative of this area is the 'JEMEZ' NORTHWESTERN FRITILLARY [*Speyeria hesperis nikias*]. You can spot this widespread fritillary throughout the Jémez uplands all summer.

Above, top to bottom:

'JEMEZ' NORTHWESTERN FRITILLARY
PALE SWALLOWTAIL

Left:

CHRYXUS ARCTIC

HISTORY HIGHLIGHT

John Woodgate

From 1905 to 1920, John Woodgate was the most important player on the New Mexico entomology and butterfly scenes. An Englishman like Cockerell, Woodgate came to the United States in 1889 at age 29. Though his early years in this country remain a mystery, he somehow ended up in northern New Mexico. Woodgate became a fence rider for local ranchers, and he carried a bug net on his rounds.

This cowboy naturalist made important insect collections in the Zuni Mountains between 1906 and 1911. Starting in 1912, Woodgate went on to collect butterflies more intensively in the Jémez Mountains. From his home in Jémez Springs, he collected insects for sale to museums in the eastern United States. Woodgate's collection of nearly 100 butterfly species from the region constituted the most detailed butterfly knowledge of any place in New Mexico at the time. Apache Skipper and the 'Jémez' Northwestern Fritillary were described from his collections. Woodgate's insect specimens were on display under glass at a bar in Jémez Springs as recently as 1970.

Above, top to bottom:

SILVER-BORDERED FRITILLARY
COMMON ROADSIDE-SKIPPER

Opposite page:

SYLVAN HAIRSTREAK

Find time to visit the Río Cebolla in July. Public access is available at Fenton Lake State Park and Seven Springs Fish Hatchery, both owned by the New Mexico Department of Game and Fish. Río Cebolla's marshy edges make it suitable for animals that need such scarce habitat. Watch for the SILVER-BORDERED FRITILLARY [*Boloria selene*] that gallivants about the wet meadows and nectars at flowers. This butterfly also flies in similar habitats in the Valles Caldera National Preserve. Sharper eyes are needed to see and identify COMMON ROADSIDE-SKIPPER [*Amblyscirtes vialis*]. Adults fly in grassy openings and draws in oak/pine woodlands, especially along streams with broadleaf grasses.

The Río Chama divides the Jémez Caldera from the Tusas Mountains to the north. The Bureau of Land Management (BLM) provides access to the Río Chama canyon northwest of Abiquiú Lake and west of Ghost Ranch. Famed artist Georgia O'Keeffe painted this stunningly beautiful landscape. Butterflies, though less famous, also give life and color to the area. Look in July along stream banks with abundant coyote willow [*Salix exigua*]. By scouting adjacent flowers, such as thistles, you should glimpse SYLVAN HAIRSTREAKS [*Satyrium sylvinus*], whose caterpillars eat the willows.

The Tusas Mountains create a broad, high plateau crossed by U.S. 64, which provides public access to the Carson National Forest between Tierra Amarilla and Tres Piedras. Good butterfly watching abounds throughout the forest, but one of the choice settings is Hopewell Lake, which rests in a mosaic of wet and dry woodlands and meadows. You'll be walking through damp grass if you visit in late June or July, and you're likely to come upon butterflies in flight as the dew evaporates in the late morning sun. Explore the damp meadows around the lake for COMMON ALPINE [*Erebia epipsodea*] and LARGE MARBLE [*Euchloe ausonides*], each dodging among tall grasses. GREENISH BLUES [*Plebejus saepiolus*] should also be there, flitting about their host clovers. QUEEN ALEXANDRA'S SULPHURS [*Colias alexandra*] may be puddling at some refreshing mud. The state's only known colony of FREIJA FRITILLARY [*Boloria freija*] was recorded at Hopewell Lake, but it has not been seen here since 1978. Search in late June and July, but also check out boggy lake margins in harder-to-reach locales.

From top:
COMMON ALPINE
LARGE MARBLE
FREIJA FRITILLARY

Expand your search radius from Hopewell Lake to the area north of U.S. 64, which has slightly higher, drier, rockier ground. Look among slopes strewn with rock jasmine for ARCTIC BLUE [*Plebejus glandon*]. Grassy areas will reveal NEVADA SKIPPER [*Hesperia nevada*] and DRACO SKIPPER [*Polites draco*], which nectar at legumes and other alpine flowers.

Above, top to bottom:

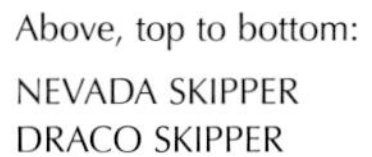

NEVADA SKIPPER
DRACO SKIPPER

Above, top to bottom:

GREENISH BLUE
QUEEN ALEXANDRA'S SULPHUR

Center:

ARCTIC BLUE

The New Mexico Department of Game and Fish operates the Sargent Wildlife Management Area on the north side of Chama. The area yields tiny, green SHERIDAN'S HAIRSTREAK [*Callophrys sheridanii*] in early May. Observe the butterflies as they nectar on dandelions and sip moisture from wet soil. Come back to the Sargent in July when the summer suite of butterflies includes the striking BLUE COPPER [*Lycaena heteronea*]. This colorful insect flies among redroot buckwheat [*Eriogonum racemosum*], its host, in the almost dry meadow just north of the main parking area. The butterfly's vibrant cousin, RUDDY COPPER [*Lycaena rubidus*], flies in damper habitats along the Chama River where its host plants, docks [*Rumex*], grow. Tiny RUSSET SKIPPERLING [*Piruna pirus*] nectars at cutleaf coneflower and puddles at streamside mud near broadleaf grasses favored by its caterpillars.

The north-central mountains are also home to small but elegant hairstreaks that fly in July and live where Gambel oak and chokecherry dominate the local landscape. You can reliably see three of these species at the Humphries Wildlife Management Area along U.S. 64, 10 miles west of Chama. BANDED HAIRSTREAKS [*Satyrium calanus*] perch on head-high branches and come to nectar. Larvae eat foliage of Gambel oak. Chokecherry [*Prunus virginiana*] hosts two hairstreaks.

From top:
BLUE COPPER
RUSSET SKIPPERLING

CORAL HAIRSTREAKS [*Satyrium titus*] nectar avidly and may hilltop. Friendly ants appear to be vital for the survival of Corals. STRIPED HAIRSTREAK [*Satyrium liparops*] adults stay near the chokecherry, perch on its leaves, and nectar at nearby flowers.

In late July and early August the north-facing slopes of the Humphries offer reliable settings to glimpse WOODLAND SKIPPER [*Ochlodes sylvanoides*] sipping nectar at meadow edges or moisture from puddles.

Above, top to bottom:

BANDED HAIRSTREAK
SHERIDAN'S HAIRSTREAK
WOODLAND SKIPPER

Center:

RUDDY COPPER

From top:

STRIPED HAIRSTREAK
CORAL HAIRSTREAK

ECO-NOTES

Myrmecophily

Ants are major invertebrate predators in terrestrial ecosystems. Lepidoptera larvae—juicy, slow-moving, and largely defenseless—would seem to be likely targets of hungry ants. Coppers, blues, hairstreaks, and metalmarks, however, have specialized physiological and behavioral adaptations that not only neutralize this threat but also convert it to an advantage. These small butterflies have mutualistic relationships with ants.

Just as ants "farm" aphids for their sugary secretions, they do the same with certain butterfly larvae. Larvae provide secretions high in amino acids and carbohydrates, which ants value highly. In return, ants detect and discourage parasitic wasps or flies by walking on or near the caterpillars and sweeping them with their antennae. Some larvae actually pupate in ant nests. In extreme cases, ants feed their young to butterfly larvae. This 15-million-year-old relationship between butterflies and ants stretches the limits of ecosystems and food webs, demonstrating the interconnectedness of the natural world.

Look for purple thistles or spiked gayfeather [*Liatris spicata*], and you'll probably spot GREAT SPANGLED FRITILLARY [*Speyeria cybele*]. This species is grander and more familiar in the eastern United States, but the more diminutive western form is no less beautiful. The latter version was described in 1876 from specimens collected in New Mexico by Lieutenant William C. Carpenter of the Wheeler Expedition.

GREAT SPANGLED FRITILLARY

HISTORY HIGHLIGHT

Military Survey Expeditions

In the early nineteenth century the U.S. government initiated an era of land acquisition in the West. Efforts began in 1803 with the Louisiana Purchase, which included the Canadian River of northeastern New Mexico. Upon signing the Treaty of Guadalupe Hidalgo in 1848, Mexico ceded to the United States most of what is now New Mexico. The remainder of the state was acquired from Mexico in 1854 through the Gadsden Purchase, after which the U.S. government established the New Mexico Territory.

The government then sponsored military-led expeditions to explore, survey, and map the new lands. Naturalists assigned to these expeditions made many scientific collections of plants, animals, and minerals in the West. Several expeditions explored New Mexico from 1820 to 1855. The names of naturalists James W. Abert, Amiel W. Whipple, and William H. Emory are familiar to most New Mexico biologists through plants (Emory oak and Whipple's penstemon) or animals (Abert's squirrel) discovered by them or named in their honor. Unfortunately, they seem to have left no butterfly specimens.

The Wheeler expeditions of 1871 to 1874 covered several western states and left the oldest traceable records of butterflies from New Mexico. Expedition naturalists

Monument to Wheeler Expedition at Wheeler Peak

Theodore L. Mead, Henry W. Henshaw, and William C. Carpenter used frontier military outposts as bases of operation. Captured butterflies were sent for identification to William H. Edwards at the Smithsonian Institution.

Expedition naturalists provided New Mexico specimens that were used to describe several butterflies new to science. They included Saltbush Sootywing and Juniper Hairstreak, the latter captured near Fort Wingate in 1874. Reports from these expeditions also included Old World Swallowtail, Black Swallowtail, and Dotted Checkerspot, the latter taken near Santa Fe.

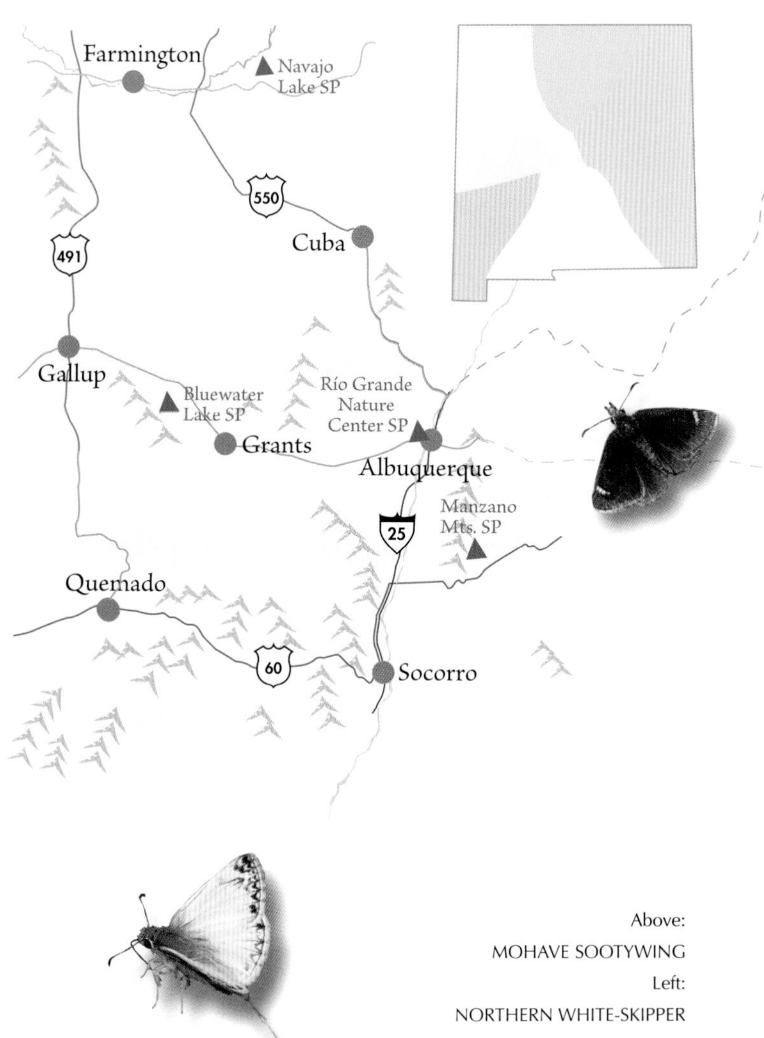

Above:
MOHAVE SOOTYWING
Left:
NORTHERN WHITE-SKIPPER

Great Basin and Colorado Plateau ecoregion butterflies prevail in semi-arid northwestern New Mexico. Butterflies of the Zuni and Chuska mountains are well known. The expansive, forbidding lowland and mesa country of the San Juan Basin, however, still holds some surprises for butterflyers. The Northwest Plateau region contains habitats associated with Upper Sonoran, Transition, and Canadian life zones.

The descent from the Continental Divide west of Chama into the San Juan Basin is long and gradual, but the aridity comes on with stunning suddenness. The surrounding mountains must grab most of whatever rain and snow fall in the region, leaving little for the dusty lowlands. This area does not contain many superb locales for butterflies, so a fair amount of scouting and exploring is necessary.

The San Juan River below Navajo Dam is a world-class trout fishery, which explains the cheek-by-jowl crowds prevalent in summer months. SANDHILL SKIPPER [*Polites sabuleti*] frequents riverside pullouts in June and again in August and September. Adults perch in gullies and swales and along river floodplains. Scruffier, drier lowlands along the San Juan River and tributary creeks and arroyos sometimes produce MOHAVE SOOTYWING [*Hesperopsis libya*] and NORTHERN WHITE-SKIPPER [*Heliopetes ericetorum*].

Left to right:
BECKER'S WHITE
SANDHILL SKIPPER
Below:
JUBA SKIPPER

These two species are more routine to the west and south at lower elevations, but they do show up here on rare occasions.

A late May trip to Jackson Lake (operated by the New Mexico Department of Game and Fish) north of Farmington provides access to some low-elevation grasslands. Ignore the lake and explore the surrounding terrain, where JUBA SKIPPER [*Hesperia juba*] perches and patrols in arroyos. Because its host mustard, prince's plume [*Stanleya pinnata*], grows widely in the region, BECKER'S WHITE [*Pontia beckerii*] can be encountered almost anywhere, often at nectar in open areas and drainages. Expect adults in May and August.

HISTORY HIGHLIGHT

Butterfly Anthropology

Images of butterflies routinely embellish books, movies, advertisements, and other contemporary visual media, but the animals themselves have been appreciated in the Southwest for eons. Nearly all cultures, modern and prehistoric, held butterflies in some regard. For Native American peoples, this relationship is expressed in pottery, rock art, architecture, and rituals.

Reverence for butterflies apparently arose long ago in Tewa culture along the Río Grande in New Mexico and to the west in Zuni lands, where many traditional art forms feature butterflies, even today. Butterflies also figured prominently in prehistoric Hopi myth and ritual. Their images occur on ancient Hopi pottery; tribal members perform a ritual butterfly dance in summer; and one Hopi pueblo has a butterfly clan. Spirit of Butterfly is personified in Zuni and Hopi dancers and katsinas. In these cultures, butterflies seem to be associated with summer, rain, fertility, and successful harvests.

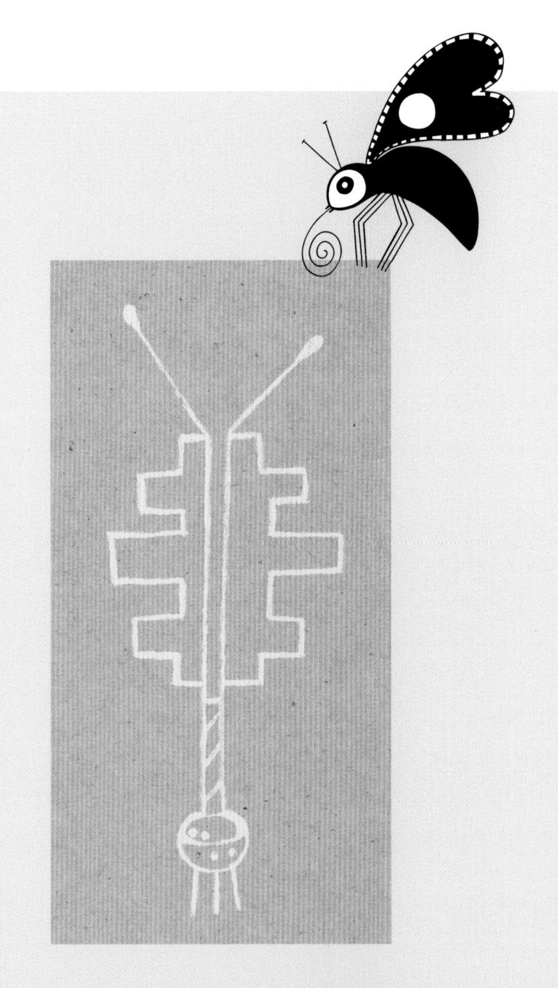

The volcanic Taos Plateau offers several rewarding butterfly destinations, including the Río Grande Wild and Scenic River and surrounding terrain. Access is north and west from Questa, through the small community of Cerro and all the way to the rim, where BLM recreation sites offer jumping-off points into the Río Grande's gaping gorge. The occasionally moderate-to-steep trail leading down into the canyon in late July or August should reveal many butterflies, including TAILED COPPER [*Lycaena arota*], which lives off the gooseberries [*Ribes*] growing along slopes and side canyons. Look for adults at moist soil or sipping nectar at asters, coneflowers, or wild buckwheat. The highlight of the excursion is YUMA SKIPPER [*Ochlodes yuma*]. This butterfly is always associated with common reed [*Phragmites australis*], a tall, emergent aquatic grass that grows at springs and seeps—scarce and vulnerable habitats in the semi-arid Southwest. To see Yuma, walk to Big Arsenic Springs, where August adults dodge among reeds and nectar at thistles.

Steve Larese

Right, top to bottom:
YUMA SKIPPER
TAILED COPPER

Uplands surrounding the San Juan Basin are islands of diversity. The Chuska Mountains, the state's most northwestern upland, harbor a few butterflies that are more common farther west but not viewable elsewhere in New Mexico. The Chuskas are part of the Navajo Nation, which has its own rules regarding access, wildlife disturbance, and related issues. The Navajo Nation also provides public parks with camping, hiking, and fishing opportunities.

If you visit the Chuskas, keep an eye out for TAWNY CRESCENT [*Phyciodes batesii*], which lives in fields and open areas at elevations of 6,500 feet to 9,000 feet. Adults are on the wing in July and come to nectar. The mountains' grassy western footslopes are the eastern edge of the range for the GREAT BASIN WOOD-NYMPH [*Cercyonis sthenele*]. Adults are most abundant in August, usually in areas ranging from 5,500 feet to 7,200 feet in elevation. The Chuskas are also home to 'CHUSKA MOUNTAINS' ANICIA CHECKERSPOT [*Euphydryas anicia chuskae*] (not shown), which has a yellowish cast and flies in July. A great deal remains to be learned about this butterfly.

Left, top to bottom:
TAWNY CRESCENT
GREAT BASIN WOOD-NYMPH
ARROWHEAD BLUE

Continue your exploration of Great Basin butterflies in the Zuni Mountains managed by the Cibola National Forest. Most of this upland is cloaked in piñon and ponderosa pine savannas. Vehicle access is available to many sites, though high-clearance and four-wheel-drive vehicles may occasionally be necessary. Rain can turn dirt roads to glue in the western part of the range.

One access point to the Zunis is via Wingate, or old Fort Wingate, south of I-40. Along with routine Transition Zone pine savanna butterflies, you can also glimpse ARROWHEAD BLUE [*Glaucopsyche piasus*] in May and June. Adults perch on and fly about patches of silver lupine [*Lupinus argenteus*], their favorite caterpillar food. A little later in summer, for example at Grasshopper Spring, watch for BEHR'S HAIRSTREAK [*Satyrium behrii*]. Adults fly in July and nectar enthusiastically, preferring small, white flowers and never straying far from their caterpillar food: mountain mahogany [*Cercocarpus montanus*].

BEHR'S HAIRSTREAK

DESERT ELFIN

A little farther east along the north flank of the Zunis, back toward Grants, you can find forest road access to Pole and Prop canyons. Here you will discover stands of cliffrose [*Cowania mexicana*] (not on cliffs, fortunately) and populations of its cute little herbivore, DESERT ELFIN [*Callophrys fotis*]. Desert Elfins fly in May near aromatic flowers of the shrubby host, perching nearby or feeding at moist sand. Cliffrose also grows at Chaco Culture National Historical Park, so this butterfly should be there, too.

From top:

ELLIS' BLUE
SPALDING'S BLUE
ROCKY MOUNTAIN DOTTED-BLUE

Wild Buckwheats

Wild buckwheats [*Eriogonum*; Polygonaceae] are a complex group of small, drought-tolerant shrubs headquartered on the Colorado Plateau. Among their butterfly herbivores are several blues, hairstreaks, coppers, and metalmarks. Some of these have host relationships that are remarkably specific: one kind of butterfly requires one specific buckwheat host.

ROCKY MOUNTAIN DOTTED-BLUE [*Euphilotes ancilla*] has been found northeast of Aztec, where 5,500-foot-high mesas support the host, cushion buckwheat [*Eriogonum ovalifolium*], a groundcover with spherical, white flower clusters in spring. Adults fly in May.

SPALDING'S BLUE [*Euphilotes spaldingi*] likes high grasslands and pine savannas with stands of its host, redroot buckwheat [*Eriogonum racemosum*]. Adults fly mid-July to mid-August, frequenting host flower clusters, nectar, and mud left by summer rains. Seek them in the Zuni Mountains, on Mount Taylor, or in the Humphries Wildlife Management Area west of Chama.

If you drive along N.M. 509 heading northwest of Grants in August you'll encounter blooming roadside stands of various wild buckwheats. Watch for MORMON METALMARK [*Apodemia mormo*]. Its caterpillars eat many varieties of buckwheat that bloom in late summer. A greater challenge on this road trip will be locating ELLIS' BLUE [*Euphilotes ellisii*], which flies only in stands of its single host, buckwheat brush [*Eriogonum corymbosum*].

MORMON METALMARK

Hilltops are worth exploring in the northwest quadrant, but prominent summits with habitat and public access are not easy to find. Mount Taylor is too high to attract species of lower elevations, but Mount Sedgwick, the highest point in the Zuni Mountains at 9,100 feet, is close to perfect. High-clearance vehicles are advisable on the forest road to the top.

You may spot various blackish swallowtails careening about the breezy summit in search of mates. OLD WORLD SWALLOWTAIL [*Papilio machaon*] flies in Transition Zone savannas in this region of the state. Old Worlds come in two forms: the black form dominates in New Mexico, but the yellow form sometimes turns up. Old World adults are prevalent in May and July. ANISE SWALLOWTAIL [*Papilio zelicaon*] has black and yellow forms, too, but the yellow version dominates here. Anise prefers Canadian Zone open conifer woodlands, and adults are visible in May and June. Anise larvae eat plants in the carrot family [Apiaceae]. Distinguishing the various blackish swallowtails in New Mexico requires careful study of minute differences.

Above:
OLD WORLD SWALLOWTAIL

Another blackish swallowtail, INDRA SWALLOWTAIL [*Papilio indra*] (not shown), also warrants mention here. It occupies some of the most starkly beautiful and least accessible lands in the Southwest, including Mesa Verde National Park in southwest Colorado. Adults fly in May. Indra seems very rare in northwest New Mexico. It was recorded in the region by John Woodgate in 1917 but has not been seen since. Surveys of the Chuska and Zuni mountains failed to locate Indra. Because of limited verified sightings, its status in New Mexico is not clear.

Turning southward, take time to explore the plateaus and mountains of west-central New Mexico—the Mogollón Highlands ecoregion. The Gila National Forest encompasses important areas where habitats remain largely undisturbed. This region has benefited from forest-management policies that retain fire as an active force in local ecosystems; hence, grassy areas and grass-dependent butterflies are abundant. Traverse N.M. 32 from Quemado to Apache Junction in June and you can see the 'MOGOLLON' COMMON RINGLET [*Coenonympha tullia subfusca*] at its easternmost occurrence. This race is more striking than the subtly shaded race of northern New Mexico. Come back in late July

From top:
ANISE SWALLOWTAIL
'MOGOLLON' COMMON RINGLET

From top:
MEAD'S WOOD-NYMPH
SNOW'S SKIPPER

and August for SNOW'S SKIPPER [*Paratrytone snowi*]. This skipper perches along dry, grassy gulches and nectars at pink and purple flowers in sunny ponderosa pine openings. Also in July, but more challenging to find, is MEAD'S WOOD-NYMPH [*Cercyonis meadii*]. We know its habitat—piñon/juniper savannas with lots of blue grama grass—but it is never seen in large numbers.

Most high-altitude butterflies of the Mogollón Highlands are closely related to species that also fly in the north-central mountains and were discussed earlier. They have evolved on separate paths for the past 10,000 years and look somewhat different. To see many of these local subspecies, drive up from Glenwood through Mogollón and into the Gila National Forest along the Bursum Road (N.M. 159), which provides access to Willow Creek Campground and Bearwallow Mountain. Middle and late summer should reveal local races of several butterflies (not shown): NORTHWESTERN FRITILLARY [*Speyeria hesperis nausicaa*], COLORADO BRANDED SKIPPER [*Hesperia colorado susanae*], QUEEN ALEXANDRA'S SULPHUR [*Colias alexandra apache*], CLOVER BLUE [*Plebejus saepiolus gertschi*], SILVERY BLUE [*Glaucopsyche lygdamus arizonensis*], SILVER-SPOTTED SKIPPER [*Epargyreus clarus huachuca*], and BOISDUVAL'S BLUE [*Plebejus icarioides buchholzi*].

Predators

Most animals find themselves on the wrong end of the predator-prey dynamic at some time in their lives. In that interaction, butterflies always are prey. This reality may clash with the conventional image of butterflies as carefree creatures of elegance, grace, and beauty. To birds and many insects, butterflies are snacks. Each stage of the butterfly life cycle is vulnerable to its own set of predators.

Caterpillars seem designed to be eaten: they are plump packets of protein that cannot escape. Predators simply have to find them, but caterpillars make that task harder through use of excellent camouflage or warning colors. Despite these and other clever defensive adaptations, moth and butterfly larvae are routinely chomped by birds. Females of certain wasps and flies place their eggs on Lepidoptera larvae so their young can nibble on caterpillar insides. To guard against these attacks, many larvae have spiky defensive structures that discourage such insects from landing. The mutualistic relationship between larvae and normally predatory ants also helps keep parasites at bay. (See the related sidebar on page 92.)

Adult butterflies have to contend with other kinds of predators. Crab spiders, ambush bugs, and preying mantises prepare ambushes at flowers and are rewarded with regular butterfly meals. Butterflies in flight are somewhat vulnerable to birds and robberflies—among the most agile and acrobatic flying insects. Butterflies may seem safe as they perch in trees, but birds also spend time there. When you see four wings scattered on the ground, you know a bird captured a butterfly, plucked off and discarded the four tasteless appendages, and gulped down the good parts.

But butterflies are not without defenses. Many tree-perching varieties of hairstreaks and swallowtails have color patterns, wing shapes, and behaviors that minimize bird predation. Undersides of these butterflies are often dull, with "false head" markings that direct a bird's attention to the back of the wings. When the butterfly slides its hind wings back and forth, its tails wiggle in the wind like antennae. If a bird strikes the bait, the backs of the wings tear away and the butterfly escapes (perhaps with a pie slice missing from its wings) to continue its search for mates.

REGIONAL SPECIALTIES

Southwest Basin and Range

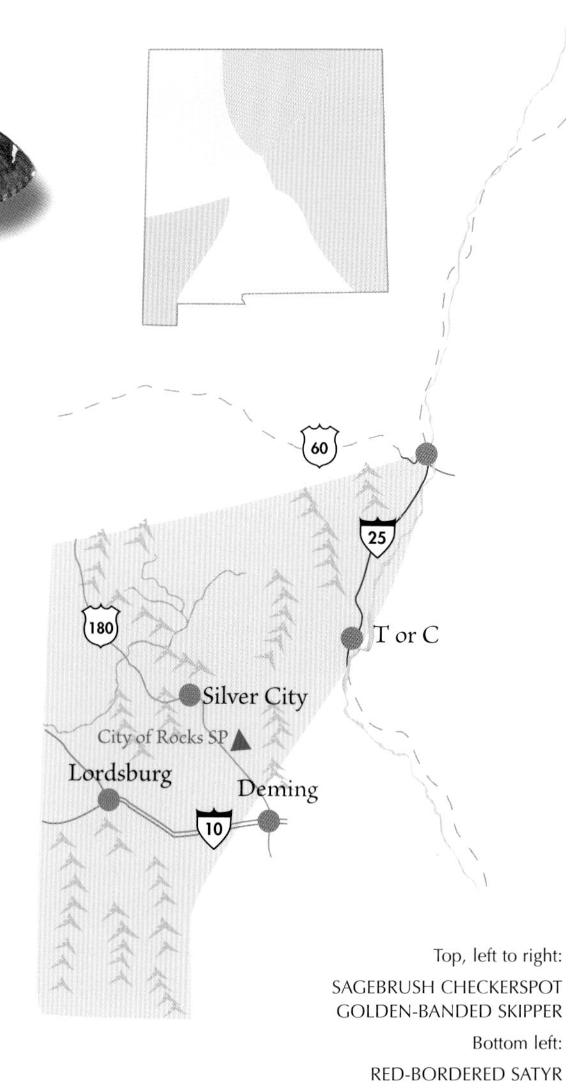

Top, left to right:

SAGEBRUSH CHECKERSPOT
GOLDEN-BANDED SKIPPER

Bottom left:

RED-BORDERED SATYR

Opposite page, bottom:

BROWN ELFIN

Southwestern New Mexico experiences the juncture of three ecoregions: the Mogollón Highlands, the Sierra Madre Uplands, and the Chihuahuan Plateau. This maelstrom of colliding ecosystems is one of the most prolific areas for butterflies in the United States. Furthermore, an annual summer influx of subtropical strays supplements resident butterfly diversity in the southern counties. The Southwest Basin and Range region can be flush with such vacationers after a good summer rainy season.

To make the most of the transition from the Gila high country, visit the Piños Altos Mountains. N.M. 15 provides access from Silver City to Gila National Forest recreation areas such as Cherry Creek and McMillan campgrounds. Come in May to see DEVA SKIPPER [*Atrytonopsis deva*], which flies, perches, and nectars in canyons within piñon/juniper/oak savanna habitats. Return during July and August to find stunning GOLDEN-BANDED SKIPPERS [*Autochton cellus*]. Males perch on shrubs along creeks and search drainage corridors for females. A visit in September should net the RED-BORDERED SATYR [*Gyrocheilus patrobas*], whose adults bob and weave deceptively among grasses and shrubs and sometimes come to nectar and wet sand.

You can spot the following three species on an April trip along N.M. 152 west of Emory Pass in Grant

DEVA SKIPPER

DESERT MARBLE

County. Stop at established Gila National Forest recreation areas such as Lower Gallinas Campground and look for BROWN ELFIN [*Callophrys augustinus*]. Adults bask and perch on host manzanita shrubs [*Arctostaphylos pungens*] or visit nectar and wet sand. SAGEBRUSH CHECKERSPOT [*Chlosyne acastus*] should also be there. Its larvae eat flowers from the daisy family, with an apparent preference for sticky aster [*Machaeranthera*]. DESERT MARBLE [*Euchloe lotta*] may be more difficult to find because it searches for mates on hilltops, and hilltops are hard to reach along this highway. Females oviposit on mustards.

For the next two butterflies, seek hilltops between 7,000 feet and 9,000 feet in elevation. Travel along

Winslow Howard

HISTORY HIGHLIGHT

A New Englander by birth, Winslow J. Howard was a Tiffany's-trained jeweler and a student of natural history. Howard traveled from New York via the Santa Fe Trail to New Mexico's territorial capitol in 1858, where he set up business as a watchmaker and jeweler. His sojourn in Santa Fe was cut short by the discovery of gold to the north, and by 1861 he was in the Colorado gold fields. For the next 30 years he traveled from one mining boomtown to another throughout the West, running an assay/jewelry business and collecting natural history specimens.

Howard eventually landed in Silver City, New Mexico, where he operated his business from 1880 to 1887. Respected as the town's jeweler, he was occasionally featured in local newspapers because of his collections of coins and natural history specimens, displayed at his shop in his "cabinet of curiosities."

While in Silver City, Howard collected butterflies and moths from the area. In 1884, entomologist Francis Snow (see page 49) came to visit. Specimens collected by Howard and Snow are the earliest documentation of the rich butterfly fauna around Silver City. The two collectors capitalized on the city's newfangled electric streetlights to capture the beautiful Northern Giant Flag Moth [*Dysschema howardi*], which was later named in Howard's honor.

Howard's interests ranged from minerals, metallurgy, and coins to plants and insects. Though not a preeminent scientist, he was one of a myriad of lesser-known collectors who provided raw material for scientific discovery in the West during the nineteenth century.

Winslow J. Howard

No photo of Winslow Howard is known, but he was a prolific letter writer. This signature appeared on an 1879 letter he wrote to lepidopterist William H. Edwards.

N.M. 152 from Hillsboro west to Emory Pass. ARIZONA HAIRSTREAK [*Erora quaderna*] adults are visible in April and July. Males stay on hilltops, and you can easily reach a few as you climb up from Kingston. Females feed at flowers and wet sand along Percha Creek above Kingston. A return to the same hills in September should afford a glimpse of APACHE SKIPPERS [*Hesperia woodgatei*]. Adults nectar at redolent rabbitbrush [*Chrysothamnus*]. The scientific name honors John Woodgate, who collected the first specimens in the Jémez Mountains in 1913. The possibility of sighting an Apache Skipper is an excellent reason to venture forth late in the season and extend your butterfly summer.

South from Silver City, N.M. 90 provides access to the Burro Mountains and the southern outpost of the Gila National Forest. Go in late May or early June to shake shrub live oaks [*Quercus turbinella*] for ILAVIA HAIRSTREAK [*Satyrium ilavia*]. This scarce Mogollón Rim foothills specialty is most likely to be nectaring near or perching on its host oaks.

Another prized find in this region is XAMI HAIRSTREAK [*Callophrys xami*]. This subtropical bug is routine in Mexico, and a few colonies exist in southeast Arizona. The single known New Mexico colony is in the Apache Box, a vertical notch cut by Apache Creek in western Grant County. The cliffs are home to two rare stonecrops

Top, left to right:
ARIZONA HAIRSTREAK
APACHE SKIPPER
Below, top to bottom:
ILAVIA HAIRSTREAK
XAMI HAIRSTREAK

From top:
DESERT ORANGETIP
'HERMOSA' ANICIA CHECKERSPOT
WHITE-BARRED SKIPPER

[*Crassulaceae*] that host Xami. Adults fly in April, July, and October and come down to nectar. You can best access Apache Box from Duncan, Arizona, via some rough roads.

A spring trip to southwestern New Mexico after normal winter rains often produces plentiful butterfly sightings. By mid-March, valley flats and low hills are freckled with spring-blooming annuals like Mexican poppies. Several butterflies target colorful mustards and other plants that are luscious green and ready for larvae to eat. Travel to low and moderate hilltops, such as those at Rockhound State Park, to see DESERT ORANGETIP [*Anthocharis cethura*]. 'HERMOSA' ANICIA CHECKERSPOT [*Euphydryas anicia hermosa*] also flies at this time, but you have to journey to the Animas Valley to find it. WHITE-BARRED SKIPPER [*Atrytonopsis pittacus*] animates these same southwestern hilltops in April and early May.

Few would consider heading deep into southwest New Mexico in July or August, but it's actually the best time to go. Monsoon season brings daily cloud buildups and frequent afternoon and evening rain showers that create mild temperatures. Creeks flow, and the landscape is as green as it ever gets. Many butterflies (and other insects) are out and about, looking for mates and laying eggs. If the rains have been good, a trip is bound to be rewarding. But beware of flash-flood possibilities and use a high-clearance vehicle.

Many of the state's southwestern mountain ranges are worth visiting during these summer months. The discussion below focuses on the Peloncillo Mountains because they offer good public access and the best menu of butterflies not usually visible in most of the United States. The only downside is that the Peloncillos are a long drive from everywhere except southeastern Arizona.

When you make the trip, be sure to search different landscape settings. Low-elevation foothill washes throughout the southern part of this region provide havens for EMPRESS LEILIA [*Asterocampa leilia*], which flies near its host, desert hackberry [*Celtis pallida*], in May and June and again in September and October. SUNRISE SKIPPER [*Adopaeoides prittwitzi*], the state's rarest tiny orange, is also a habitat specialist. Colonies occur only in grassy, slack-water habitats along lower reaches of Cloverdale Creek and Clanton Draw in the Animas Valley—on private land, as luck would have it. There appear to be two flights: May through June and August through September. Meanwhile, hilltops west of Geronimo Pass may yield 'ARIZONA' JUVENAL'S DUSKYWING [*Erynnis juvenalis clitus*], which lives on the many scrubby oaks in this area.

After the obligatory hilltop visit, return to drainages like Clanton Draw, Skeleton Canyon, and Cottonwood Canyon. You'll encounter many different butterflies searching stream corridors for mates, balancing electrolytes at

From top:
EMPRESS LEILIA
'ARIZONA' JUVENAL'S DUSKYWING
SUNRISE SKIPPER

Evergreen Oaks

From New Mexico's northern border and heading toward the equator, oaks *[Quercus]* diversify from one or two species to several and shift from deciduous (losing leaves in winter) to evergreen (retaining green leaves year-round). Oaks that cloak our southern, and especially southwestern, desert mountains include Arizona white oak, Emory oak, gray oak, silverleaf oak, sandpaper oak, netleaf oak, and Toumey oak. Oak-studded landscapes in the New Mexico Bootheel provide fertile habitat for numerous butterflies whose caterpillars eat oak foliage.

Among the many Bootheel butterflies that rely on these plants for caterpillar food are ZELA METALMARK [*Emesis zela*], whose adults are on the wing in April and August, and the very similar ARES METALMARK [*Emesis ares*], which flies only in late summer. Both metalmarks are attracted to flowers that grow in oak-bordered ravines and creek beds. These same late-summer flowers are magnets for the striking DULL FIRETIP [*Apyrrothrix araxes*], which can be confused with no other skipper in New Mexico. Beyond the three species presented here, oak-based butterflies probably comprise 10 percent of all butterfly species in this corner of New Mexico.

From top:
ZELA METALMARK
ARES METALMARK
DULL FIRETIP

puddles, or nectaring at late-summer flowers. Look for acacia-feeders ACACIA SKIPPER [*Cogia hippalus*] and CAICUS SKIPPER [*Cogia caicus*], which are fairly common in the southwest quadrant. They bounce along arroyo bottoms looking for mates, nectar, and wet soil.

If it's your first trip to this country, the abundant, lush grasses of late summer will amaze you. Not surprisingly, many Lepidoptera herbivores are here to capitalize on the cornucopia. Summer-flying local butterflies with grass-feeding caterpillars include MOON-MARKED SKIPPER [*Atrytonopsis lunus*], LARGE ROADSIDE-SKIPPER [*Amblyscirtes exoteria*], and SLATY ROADSIDE-SKIPPER [*Amblyscirtes nereus*].

Top row, left to right:

SLATY ROADSIDE-SKIPPER
CAICUS SKIPPER
LARGE ROADSIDE-SKIPPER

Middle:

MOON-MARKED SKIPPER

Bottom:

ACACIA SKIPPER

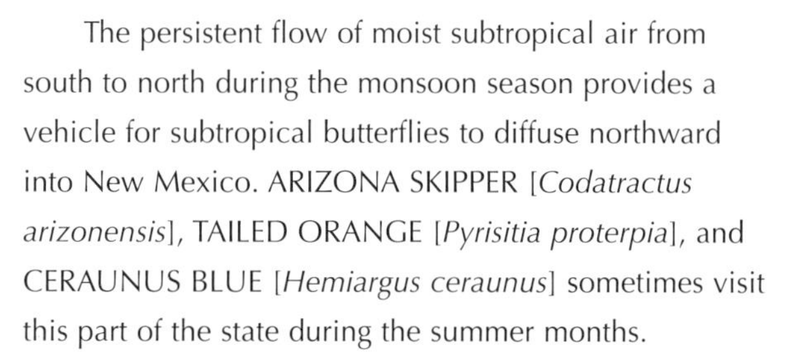

The persistent flow of moist subtropical air from south to north during the monsoon season provides a vehicle for subtropical butterflies to diffuse northward into New Mexico. ARIZONA SKIPPER [*Codatractus arizonensis*], TAILED ORANGE [*Pyrisitia proterpia*], and CERAUNUS BLUE [*Hemiargus ceraunus*] sometimes visit this part of the state during the summer months.

The region may exhibit its major butterfly show in July and August, but additional local specialties continue to emerge from chrysalids during the gradual post-summer slide into autumn. Come back in September or October to see DESERT CHECKERED-SKIPPER [*Pyrgus philetas*] on footslopes of desert mountains and in nearby arroyos. It resembles the White and Common Checkered-Skippers, but its flight is more subdued. ARIZONA METALMARK [*Calephelis arizonensis*] is a rare arroyo-dweller at mid-range elevations.

Above, top to bottom:

ARIZONA SKIPPER
CERAUNUS BLUE
DESERT CHECKERED-SKIPPER

Across, left to right:

ARIZONA METALMARK
TAILED ORANGE

As you drive up Clanton Draw in October, ARIZONA GIANT-SKIPPERS [*Agathymus aryxna*] buzz noisily around canyon mudholes. Go slowly so they have time to get out of your way. Nearby rocky hills support Palmer's agave [*Agave palmerii*], their preferred caterpillar food. Higher country in the Peloncillos occasionally reveals CHIRICAHUA WHITE [*Neophasia terlooii*], whose white males float amidst the pine canopy, descending in search of orange females. The species is more common in pine forests in the Mexican Sierra Madre and in the Chiricahua Mountains of southeastern Arizona.

From top:
ARIZONA GIANT-SKIPPER
CHIRICAHUA WHITE

From top:
ERICHSON'S WHITE-SKIPPER
DESERT CLOUDYWING
UMBER SKIPPER

Subtropical strays that rarely pass through southwestern New Mexico include the following species. You can sometimes spot them at moisture or nectar in well-watered canyons: DESERT CLOUDYWING [*Achalarus casica*], ERICHSON'S WHITE-SKIPPER [*Heliopyrgus domicella*], YELLOW ANGLED-SULPHUR [*Anteos maerula*], RUDDY DAGGERWING [*Marpesia petreus*], TROPICAL CHECKERED-SKIPPER [*Pyrgus oileus*], and UMBER SKIPPER [*Poanes melane*].

Guadalupe Canyon, at the south end of the Peloncillos, opens to the southwest, connects to Mexico's Río Yaqui, and is a conduit for many subtropical butterflies diffusing north during the summer monsoon season. Its location and orientation make it the best "trap" canyon for butterfly immigrants in New Mexico. Guadalupe Canyon is accessible only by driving east from Douglas, Arizona. For more details on accessing this hard-to-reach site, see the New Mexico Ornithological Society's *New Mexico Bird Finding Guide* in the Recommended Reading section.

From top:
RUDDY DAGGERWING
TROPICAL CHECKERED-SKIPPER
YELLOW ANGLED-SULPHUR

Top, left to right:

LONG-TAILED SKIPPER
DORANTES LONGTAIL

Above:

BROWN LONGTAIL

New Mexico can claim some butterflies as resident species only because they live in Guadalupe Canyon. SHEEP SKIPPER [*Atrytonopsis edwardsi*] flies in early September. Adults perch with heads down, high on canyon walls, and come to moist earth. POLING'S GIANT-SKIPPER [*Agathymus polingi*] flies in October and also perches, often head-down in the shade, on canyon walls. Its name honors Otto C. Poling, a prolific nineteenth-century collector in the Southwest. Larvae eat Schott's agave [*Agave schottii*], a small plant that grows on canyon walls. Another autumn flyer is NABOKOV'S SATYR [*Cyllopsis pyracmon*], which dodges about on shady canyon bottoms.

The other enticement to undertake the lengthy trip to the canyon is the (still small) chance of seeing subtropical strays without leaving the United States. Special vagrants occasionally found here include the spectacular LONG-TAILED SKIPPER [*Urbanus proteus*], DORANTES LONGTAIL [*Urbanus dorantes*], BROWN LONGTAIL [*Urbanus procne*], and WHITE-STRIPED LONGTAIL [*Chioides albofasciatus*].

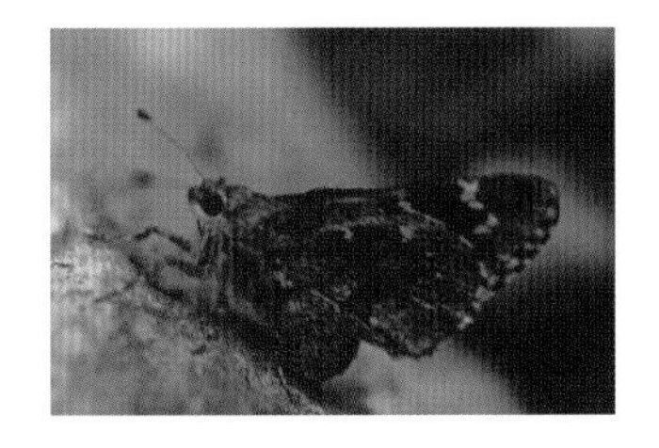

Bottom, left to right:

NABOKOV'S SATYR
SHEEP SKIPPER
POLING'S GIANT-SKIPPER

Butterflies as Pollinators

Butterflies play key roles in ecosystems as eaters of plants and food for predators. They also play a vital role as pollinators for plants. Pollination relationships between plants and insects are among the most sacred on the planet, providing food and energy for one partner and reproduction for the other. Bees are the most efficient pollinators, but butterflies are next best. Few other insects have such a strong "good cop-bad cop" (pollinator-herbivore) relationship with plants.

When we see a butterfly visit a flower, we ask questions. Why does the butterfly risk danger from predators to visit the flower? Well, the butterfly needs food and therefore siphons the nectar from the flower. How nice of the flower to provide nectar, but what's in it for the plant? Why invest a lot of water and sugar to create nectar? The plant needs the butterfly to move pollen from stamen to pistil in order to make seeds. You could say that plants invented artificial insemination. The plant does the hard work: it makes the nectar, colors its petals to advertise the sweet liquid, and arranges its parts so pollen transfer and seed formation can occur. The butterfly simply comes to eat.

Flowering plants and pollinating insects co-evolved over eons of time, while natural selection adjusted the characteristics of these partners. Each flower's size, shape, color, nectar production, and bloom time are designed to maximize the transfer of pollen. For their part, butterflies have senses, behaviors, and feeding devices designed to find and harvest nectar. The beauty of flowers and butterflies brings us pleasure, but for them it manifests nothing less than the biological imperative to survive and reproduce.

WHITE-STRIPED LONGTAIL

Above:

RED SATYR

Chihuahuan flora and fauna dominate the desert borderlands of south-central New Mexico, which stretch from Deming to Carlsbad. This area is complex; calling it a desert is an unfair stereotype. True, it is hot in summer, but it also experiences killing frosts in winter. The region includes riparian environs, pine forests in isolated mountain islands, and vast grasslands like those of the Great Plains. It also contains lowlands with annual rainfall averaging less than ten inches, sufficient to support scrubby, drought-tolerant vegetation such as creosote bush and many kinds of cactus.

Explore the area with a critical eye and an open mind. Don't be fooled into thinking that this so-called desert is bereft of life. In truth, it is one of the state's most diverse areas for plants, butterflies, and other living things. Do be cognizant that most organisms found here, including butterflies, are sensitive to moisture conditions and may be in a dormant stage if rainfall has been scarce. Watch weather patterns and time your explorations accordingly. If winter is wet, then butterflies abound in March and April; if the summer monsoon season is good, then butterflies are numerous from July through October. These are two big "ifs."

Grasslands are extensive in southern New Mexico. At elevations above 6,000 feet they are home to RED SATYR [*Megisto rubricata*]. Adults fly from June through August, dodging deliberately among grasses at locales such as the BLM's Aguirre Springs Recreation Area in the Organ Mountains northeast of Las Cruces.

Honey mesquite [*Prosopis glandulosa*] is a staple shrub of this area. Its extent and density are on the increase because cattle eat the delicious and nutritious seedpods and then "plant" the seeds a few hours later. Below 6,000 feet in elevation, honey mesquite directly supports two small butterflies whose larvae eat the plant's foliage. PALMER'S METALMARK [*Apodemia palmerii*] adults are abundant in May and June and again in August and September. LEDA MINISTREAK [*Ministrymon leda*] adults fly in June and September. Both species nectar at small flowers and can wander to nonbreeding areas.

Above, left to right:
LEDA MINISTREAK
PALMER'S METALMARK

An August-September trip to Hadley Draw in the Cooke's Range gives you a chance to glimpse an interesting critter. Rocky parts of this area contain Parry's agave [*Agave parryi*], which hosts ORANGE GIANT-SKIPPER [*Agathymus neumoegeni*]. Males patrol rocky hilltops, though both sexes visit canyon mud holes.

Speaking of hilltops, the region has many. They are particularly productive butterfly habitats, so be sure to include some high spots in your ramblings. Start with a visit to the hills located on BLM land just west of Dripping Springs in the Organ Mountains. Their height suits both butterflies and hikers. Colorful hilltop occupants in this region include two cousins. THEONA CHECKERSPOT [*Chlosyne theona*] inhabits paintbrush-studded grasslands. It takes flight from April through May and again in August. BORDERED PATCH [*Chlosyne lacinia*] also flies in spring and late summer. Its caterpillars eat sunflowers [*Helianthus*]. Adults have narrow, white bands or broad, orange bands.

Above, top to bottom:

ORANGE GIANT-SKIPPER
BORDERED PATCH

Top right:

THEONA CHECKERSPOT

Plentiful scrubby oaks contribute significantly to butterfly diversity here. Both MERIDIAN DUSKYWING [*Erynnis meridianus*] and 'ARIZONA' MOURNFUL DUSKYWING [*Erynnis tristis tatius*] have oak-feeding caterpillars. Oaks, junipers, and other deciduous trees are parasitized by broadleaf mistletoes [Viscaceae] that in turn host GREAT PURPLE HAIRSTREAK [*Atlides halesus*]. Males of the species perch atop hilltop shrubs and chase all comers.

The many low mountain ranges and volcanic hills are dissected by numerous dry arroyos and washes that are habitats for varied, small butterflies. ORANGE SKIPPERLINGS [*Copaeodes aurantiaca*] virtually own these habitats. Oblivious of their small stature, males aggressively defend territories in gullies and washes. They perch alertly and dart after all intruders, often returning to the same rock-top perch. A spring generation flies in April, and overlapping broods surge in late summer.

Top to bottom:
GREAT PURPLE HAIRSTREAK
'ARIZONA' MOURNFUL DUSKYWING
ORANGE SKIPPERLING
Right:
MERIDIAN DUSKYWING

Arroyos among grassy hills harbor CARUS SKIPPER [*Polites carus*] in May and again from July through August. TEXAS ROADSIDE-SKIPPER [*Amblyscirtes texanae*] likes rocky canyons but dwells lower in the landscape than other rocky-canyon roadside-skippers, usually below 6,000 feet. Expect adults in May and July. DOTTED ROADSIDE-SKIPPER [*Amblyscirtes eos*] is plentiful in eastern and southern New Mexico lowlands. Adults fly from April through May and again in August, preferring to perch on the ground in low spots.

Scrubby Chihuahuan washes, such as the one at Rockhound State Park near Deming, are like Grand Central Station for three skippers: GOLDEN-HEADED SCALLOPWING [*Staphylus ceos*], ARIZONA POWDERED-SKIPPER [*Systasea zampa*], and COMMON STREAKY-SKIPPER [*Celotes nessus*]. They fly near arroyo bottoms and edges in April and August, dodging through shrubs, perching on moist earth or low plants, and nectaring at flowers.

Above, top to bottom:

CARUS SKIPPER
DOTTED ROADSIDE-SKIPPER
GOLDEN-HEADED SCALLOPWING
ARIZONA POWDERED-SKIPPER

Center, from top:

COMMON STREAKY-SKIPPER
TEXAS ROADSIDE-SKIPPER

And don't forget the most adorable and smallest checkerspots: TINY CHECKERSPOT [*Dymasia dymas*] and ELADA CHECKERSPOT [*Texola elada*]. These closely related species are rarely found together, though they share scrubby desert wash habitats, low and weak flight mechanics, spring/late-summer broods, and hosts in the acanthus family.

The Gila River, from the confluence of Mogollón Creek near Cliff downstream to the Arizona border near Virden, offers several opportunities to investigate Chihuahuan riparian butterfly habitats. (Geographically this area is located in the Southwest Basin and Range region.) You can observe DESERT VICEROY [*Limenitis archippus obsoleta*] adults in May and August, perching high on host Goodding's willows [*Salix gooddingii*]. Another riverside tree, netleaf hackberry [*Celtis reticulata*], hosts TAWNY EMPEROR [*Asterocampa clyton*], which feeds on tree sap. You might also catch sight of EASTERN TAILED-BLUE [*Cupido comyntas*] in herbaceous plant communities along the river.

Clockwise from top left:
TINY CHECKERSPOT
DESERT VICEROY
EASTERN TAILED-BLUE
ELADA CHECKERSPOT
TAWNY EMPEROR

Drought and the Butterfly "Seedbank"

ECO-NOTES

Over the past 10 years, New Mexico has experienced intervals of persistent, severe drought. Butterflies are scarce during droughts, but where do they go? Will they come back?

Plants and animals use dormancy to protect themselves during predictable, seasonal dry spells and even during long, unscheduled droughts. Farmers and gardeners know that seeds kept in a cool, dry place can last a long time in a dormant state. If you put them in warm, wet ground, many will germinate even after years have passed. Many animals have similar capabilities. Bears hibernate through cold winters. Desert-dwelling spadefoot toads bury themselves underground until the next summer rainy season. Dormancy helps these organisms survive when food and water are unavailable.

Insects exhibit this behavior, too. They survive long, inhospitable periods of cold or dry conditions by entering an inactive state called diapause. One classic desert insect, the yucca moth, has eggs that can survive 20 or 30 years in diapause. Then, when conditions are right, caterpillars hatch and begin feeding. Among butterflies, extended diapause has been documented for Sara Orangetips, which normally overwinter as chrysalids and fly in spring after winter rains. In New Mexico, orangetips and other spring-flying butterflies are usually scarce or absent in springs following dry winters. Years later, however, after normal winter moisture, these species fly in regular numbers.

It seems clear that many butterflies survive drought by extending diapause while in an immature life stage. Like the patient seeds of desert plants, this butterfly "seedbank" bides its time and resumes development when conditions are suitable.

TROPICAL LEAST SKIPPER [*Ancyloxypha arene*] prefers riverbanks with broadleaf streamside grasses that feed its larvae. Adults flit weakly among the grasses along the Río Grande at Percha Dam State Park, with peak numbers in September. In somewhat drier spots at the park, you can find FATAL METALMARK [*Calephelis nemesis*] flying in June and again from August through October.

Willows that grow along streams in low mountains host a butterfly closely related to the Viceroy. 'ARIZONA' RED-SPOTTED ADMIRAL [*Limenitis arthemis arizonensis*] thrives in such habitats. Adults fly from May to October in overlapping generations at Dog Canyon in Oliver Lee Memorial State Park, where they cruise the canyons, perch on overhanging branches, and feed at red flowers.

Top, left to right:
FATAL METALMARK
TROPICAL LEAST SKIPPER
Right:
'ARIZONA' RED-SPOTTED ADMIRAL

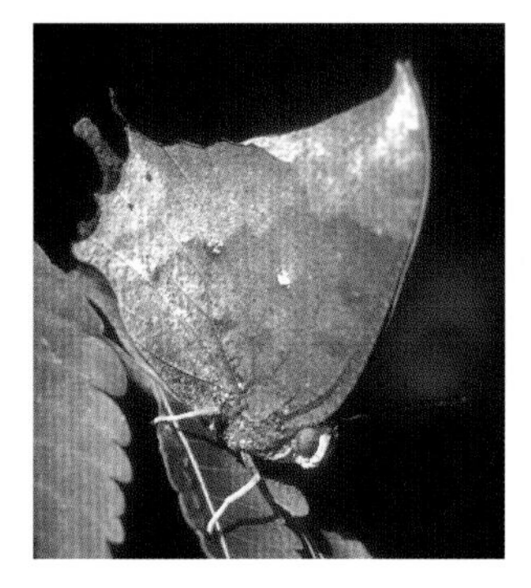

Several subtropical strays that routinely wander north from Mexico supplement the rich resident butterfly fauna in the Desert Borderlands region. These visitors are most frequent in late summer and include LARGE ORANGE SULPHUR [*Phoebis agarithe*] and GIANT SWALLOWTAIL [*Papilio cresphontes*], both of which are fond of thistle nectar. LYSIDE SULPHURS [*Kricogonia lyside*] characteristically flutter beneath low tree canopies in desert washes, fly up into the branches, and "disappear" by perching with closed wings among leaves bathed in diffuse, shaded light.

If you explore desert watercourses, you might come upon smaller subtropical vagrants that have meandered aimlessly northward. Of this crowd, TEXAN CRESCENT [*Anthanassa texana*] may be the most familiar, though TROPICAL BUCKEYE [*Junonia genoveva nigrosuffusa*] and TROPICAL LEAFWING [*Anaea aidea*] are not unexpected. Texan Crescent and Tropical Buckeye adults patrol bottoms of moist desert canyons in search of nectar, moisture, and mates.

Top row:

LARGE ORANGE SULPHUR
TEXAN CRESCENT

Middle:

TROPICAL BUCKEYE

Bottom row:

LYSIDE SULPHUR
TROPICAL LEAFWING

FIERY SKIPPER [*Hylephila phyleus*] is fairly common in New Mexico's southern towns and disturbed areas, where females find non-native lawn grasses (like Bermuda grass) on which to place eggs. Adults may even fly through mild winters in cities such as Las Cruces, Carlsbad, and Hobbs.

If you think of the resident desert borderlands butterflies as a scoop of ice cream and the routine strays as hot fudge sauce, then this next bunch is the cherry on top. These strays are indeed infrequent, but their occasional sightings are memorable. They include CYNA BLUE [*Zizula cyna*], HAMMOCK SKIPPER [*Polygonus leo*], CLOUDED SKIPPER [*Lerema accius*],

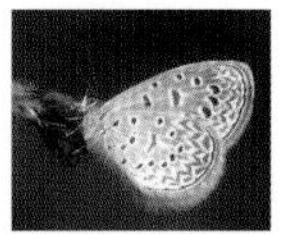

Above, left to right:
CYNA BLUE
CLOUDED SKIPPER

Left, top to bottom:
HAMMOCK SKIPPER
FIERY SKIPPER
Right:
GIANT SWALLOWTAIL

EUFALA SKIPPER [*Lerodea eufala*], ORANGE-BARRED SULPHUR [*Phoebis philea*], WHITE ANGLED-SULPHUR [*Anteos clorinde*], MIMOSA YELLOW [*Pyrisitia nise*], and BOISDUVAL'S YELLOW [*Eurema boisduvaliana*]. These critters are most often observed in low, wooded, riparian desert oases that furnish lots of nectar, water, and shelter.

The journey from Las Cruces to Carlsbad crosses the hardscrabble region known as the Trans-Pecos. Topping out at over 8,500 feet in elevation, Organ Needle and Guadalupe Peak are sentinels for the area. To the south, widely spaced upland islands leapfrog toward the Mexican Sierra Madre Oriental. To the north, the Guadalupe Mountains transition upward in big, broad shelves toward the complex of the Sacramento Mountains, Capitán Mountains, and Sierra Blanca. The southwestern part of the Great Plains ecoregion stretches eastward. The vast, low Tularosa Basin and the northern part of the Chihuahuan Plateau plunge to the west. The Trans-Pecos brings together some of the most unusual and challenging butterfly destinations in New Mexico.

Top to bottom:

MIMOSA YELLOW
ORANGE-BARRED SULPHUR
WHITE ANGLED-SULPHUR

Center top:

BOISDUVAL'S YELLOW

The Organ and Guadalupe mountains share some intriguing butterflies. For most of the warm season, DEFINITE PATCH [*Chlosyne definita*] and GRAPHIC CRESCENT [*Phyciodes graphica*] contribute to the local butterfly flavor—like the black pepper in Cajun cooking. Both feed plants in the acanthus family to their caterpillars. Seek them out in Last Chance Canyon on the way to Sitting Bull Falls.

A brief, late March trip east from Las Cruces can net DURY'S METALMARK [*Apodemia duryi*], which lives on stands of ratany [*Krameria*] on the northern footslopes of A Mountain. Continue to Dripping Springs and search adjacent canyons and arroyos for New Mexico buckeye [*Ungnadia speciosa*], a large, woody shrub whose bright pink flowers are easy to discern on early spring days. Here you will find its herbivore, HENRY'S ELFIN [*Callophrys henrici*].

Clockwise from top:
DEFINITE PATCH
EUFALA SKIPPER
HENRY'S ELFIN
DURY'S METALMARK
GRAPHIC CRESCENT

You can also visit Whites City south of Carlsbad to locate Dury's Metalmark and then continue southwest to Slaughter Canyon for Henry's Elfin. While you're in that area, continue south of Carlsbad to the Texas border. You may find TEXAS POWDERED-SKIPPER [*Systasea pulverulenta*] among hot, thorny arroyos that cross U.S. 62-180. It flies in spring and again in late summer.

Between Carlsbad and the Texas border lies Rattlesnake Springs, part of Carlsbad Caverns National Park. This desert oasis is the best locale in the region to seek subtropical strays during and after the summer monsoon season. Once vagrant butterflies arrive, they are reluctant to leave because the site offers plentiful water and abundant shelter in the form of large cottonwoods and willows. The other major attraction is nectar from extensive stands of milkweed. Odd subtropical wanderers seen here but few other places include ORNYTHION SWALLOWTAIL [*Papilio ornythion*], ZEBRA [*Heliconius charithonia*], and WHITE PEACOCK [*Anartia jatrophae*]. The acres of milkweeds also create one of the largest seasonal breeding concentrations of Monarchs and Queens in New Mexico.

From top left:
WHITE PEACOCK
ORNYTHION SWALLOWTAIL
TEXAS POWDERED-SKIPPER

The Guadalupes support two classic desert butterflies linked to the semi-arid landscape through their larval hosts. In late September and early October, walk Rattlesnake Canyon Trail in Carlsbad Caverns National Park. Watch for MARY'S GIANT-SKIPPER [*Agathymus mariae*], which perches in stands of shindagger agave [*Agave lechuguilla*]. URSINE GIANT-SKIPPER [*Megathymus ursus*] also lives here; its caterpillars bore into Torrey yucca [*Yucca torreyii*]. Finding adults, however, is quite difficult. Your best chance is to look in July on very prominent hilltops, such as Guadalupe Peak in Guadalupe Mountain National Park just across the border in Texas. If you make the long, hot, steep climb, you'll understand why this butterfly is rarely seen.

POLING'S HAIRSTREAK [*Satyrium polingi*] frequents the Organ and Guadalupe mountains and also the brushy, oak-covered, south-facing footslopes of the Capitán Mountains. Adults perch on shrub live oak hosts in late May and early June.

Top row, left to right:

POLING'S HAIRSTREAK
MARY'S GIANT-SKIPPER

Middle:

ZEBRA

Right:

URSINE GIANT-SKIPPER

Rails to the Sacramentos

HISTORY HIGHLIGHT

Transportation options controlled scientific access to New Mexico during the Frontier period. Exploration of butterflies in the Sacramento Mountains was directly related to the coming of the railroad to the region. The first flurry of butterfly study occurred in the 1890s, when botanist Elmer O. Wooton and entomologist Charles Townsend ventured there via horse and wagon from the agricultural college in Mesilla. Scientific interest in the Sacramentos intensified as transportation obstacles diminished. The El Paso and Northeastern Railroad reached Alamogordo in 1898 and completed a line up to Cloudcroft in 1899. Cloudcroft immediately became a popular tourist destination, where Alamogordo families and El Pasoans vacationed to escape summer heat. The area, however, still remained poorly accessible to scientists from elsewhere.

The situation changed in February 1902 when the Chicago, Rock Island and El Paso Railroad opened a diagonal route from Kansas across the Oklahoma and Texas panhandles to Santa Rosa, Carrizozo, and Alamogordo. Eastern scientists wasted no time. In April 1902 the Philadelphia Academy of Natural Sciences sent biologists to obtain specimens of insects, plants, mammals, birds, and reptiles from this unexplored area. Near Alamogordo, expedition biologist Henry Viereck captured a skipper that was new to science and named in his honor. Warren Knaus, an entomologist at the University of Kansas, followed months later to collect beetles and butterflies from El Paso, the Tularosa Basin, and Cloudcroft. Impressed, Knaus returned in 1903.

Butterfly specimens collected by these early rail-riding scientists still reside quietly in collections at major scientific institutions such as the Carnegie Museum, American Museum of Natural History, National Museum of Natural History (Smithsonian), and Snow Entomological Museum at the University of Kansas.

A fitting conclusion to your statewide tour of New Mexico's butterfly landscapes is an excursion to the battery of uplands that includes the Sacramento and Capitán mountains. A drive or walk in these settings reveals spruce/fir forests, alpine meadows, elk, bears, snow-capped peaks, and ski areas. You have left the plains and deserts behind, and you are not in Kansas—or Texas—anymore.

In this unique archipelago of cool, forested uplands surrounded by plains and deserts—more than 100 miles from the nearest similar cool habitat—a biologist expects to discover interesting life forms. Sure enough, these uplands contain 50 distinctive species of plants found nowhere else. Butterflies in this landscape are also special.

Drive the road to the Ski Apache ski area from the base of Sierra Blanca on a warm day in May or June, when runoff from melting snowbanks grows dandelions in sunny openings. Check the yellow flowers for little green triangles that might be 'SACRAMENTO' SHERIDAN'S HAIRSTREAK [*Callophrys sheridanii sacramento*]. This butterfly is linked to its sole larval host, Wooton's buckwheat [*Eriogonum jamesii wootoni*], which occurs throughout the Sacramentos.

Right:
'SACRAMENTO' SHERIDAN'S HAIRSTREAK

From top:
'CAPITAN MOUNTAINS' NORTHWESTERN FRITILLARY
'SACRAMENTO MOUNTAINS' ANICIA CHECKERSPOT

Cloudcroft is a good base for a July jaunt to nearby meadows. Be sure to go in the morning before the afternoon clouds roll in. Flowery subalpine meadows are heavenly in any case, but in the Sacramentos you have a chance to see some wonderful butterflies in the bargain. Look for yellow-orange flowers of sneezeweed [*Helenium*], which are popular nectar sources for many butterflies. Among the assembled hordes, pay particular attention to the tiny, coppery bronze FOUR-SPOTTED SKIPPERLING [*Piruna polingii*]. Contrast that with the larger and brighter 'CAPITAN MOUNTAINS' NORTHWESTERN FRITILLARY [*Speyeria hesperis capitanensis*], whose adults nectar in sunny meadows but whose caterpillars feed on forest floor violets.

You have to be a little more selective to find 'SACRAMENTO MOUNTAINS' ANICIA CHECKERSPOT [*Euphydryas anicia cloudcrofti*] because it does not occupy every meadow. This butterfly was discovered at the Lincoln National Forest's Pine Campground and still occurs here. Or you could visit Sleepy Grass Campground, where adults cavort lazily around their meadow homes; they seem to know that they taste bad to predators and flaunt it.

Ending your statewide sojourn in the Sacramento Mountains allows you to view butterfly ambassadors from many distinctive ecoregions and life zones found elsewhere in New Mexico. The Great Plains' Hobomok Skipper at its southwesternmost occurrence snuggles up against the Great Basin's Sagebrush Checkerspot at its easternmost occurrence. The Rocky Mountains' Tawny-edged Skipper flying at its southern limit shares cutleaf coneflower nectar with Mexico's Four-spotted Skipperling flying at its northernmost haunt. The unique juxtaposition of butterflies from distant compass points within the complex of the Sacramento uplands adds an exclamation point to your statewide butterfly tour.

FOUR-SPOTTED SKIPPERLING

ADDITIONAL RESOURCES

- ACCESSING BUTTERFLY LANDSCAPES
- CHECKLIST OF NEW MEXICO BUTTERFLIES
- BUTTERFLY GLOSSARY
- RECOMMENDED READING
- STATE PARKS
- STATE MAP
- INDEX

Above:
MALACHITE
Left:
PALAMEDES SWALLOWTAIL

Accessing Butterfly Landscapes

New Mexico offers many opportunities to observe butterflies. People wishing to access butterfly landscapes must first determine the ownership of the land to find out if any special regulations apply.

New Mexico contains large acreages of public land under several jurisdictions. Many federal agencies (Forest Service, Bureau of Land Management, Fish and Wildlife Service, Bureau of Reclamation, and Army Corps of Engineers) normally do not restrict access for butterfly observation purposes. The National Park Service operates national parks and monuments where it encourages non-consumptive butterfly appreciation. Its primary mission is to protect our natural heritage; collecting, therefore, is illegal without special permission. The Department of Defense operates several military installations, including White Sands Missile Range, and maintains very restrictive access rules. The Department of Energy operates two research laboratories in New Mexico and has similar access limits. These restrictions are in part for the public's safety.

The State of New Mexico also manages public lands. New Mexico State Parks operates 35 public parks where visitors are invited to enjoy natural resources. Nonconsumptive activities are welcome; collecting requires a permit. The Department of Game and Fish manages wildlife and fishing areas for the primary benefit of people with licenses to hunt and fish. Recognizing the growing ranks of wildlife watchers, the department recently instituted its Gain Access Into Nature (GAIN) program. Now one can purchase an inexpensive GAIN permit to secure access to wonderful butterfly habitat. The Commissioner of Public Lands, through the State Land Office, oversees state trust lands. Statute dictates management of trust lands to maximize revenue for state education, so lands are leased to income-generating activities such as grazing, mining, and logging. The State Land Office nonetheless offers recreational access to lands under its management through the purchase of a $25 annual permit.

The state has much tribal land under the jurisdiction of Native American groups. Indian reservations and other tribal lands are neither parks nor public lands; they are the lands of sovereign nations. Most tribes have specific regulations regarding access for any purpose. Restrictions on photography often apply, and sometimes the land is closed completely for tribal ceremonies. Call and inquire before visiting.

All federal, state, and tribal agencies reserve the right to restrict access in areas near species of special conservation concern. Agencies also restrict access during severe droughts when a stray spark or match could cause a major public emergency. When in doubt, call ahead.

For privately owned lands, visitors must get permission from the property owner.

Checklist of New Mexico Butterflies

Use this handy checklist to check off the New Mexico butterflies that you spot. The checklist is presented in the sequence used in J. P. Pelham's "A Catalogue of the Butterflies of the United States and Canada."

SKIPPERS (FAMILY HESPERIIDAE)

Dicot Skippers (Subfamily Eudaminae)

___ Silver-spotted Skipper (*Epargyreus clarus*)
___ Hammock Skipper (*Polygonus leo*)
___ White-striped Longtail (*Chioides albofasciatus*)
___ Zilpa Longtail (*Chioides zilpa*)
___ Short-tailed Skipper (*Zestusa dorus*)
___ Arizona Skipper (*Codatractus arizonensis*)
___ Long-tailed Skipper (*Urbanus proteus*)
___ Dorantes Longtail (*Urbanus dorantes*)
___ Brown Longtail (*Urbanus procne*)
___ Golden-banded Skipper (*Autochton cellus*)
___ Desert Cloudywing (*Achalarus casica*)
___ Drusius Cloudywing (*Thorybes drusius*)
___ Northern Cloudywing (*Thorybes pylades*)
___ Mexican Cloudywing (*Thorybes mexicana*)
___ Acacia Skipper (*Cogia hippalus*)
___ Caicus Skipper (*Cogia caicus*)

Spread-wing Skippers (Subfamily Pyrginae)

___ Dull Firetip (*Apyrrothrix araxes*)
___ Golden-headed Scallopwing (*Staphylus ceos*)
___ Common Sootywing (*Pholisora catullus*)
___ Mexican Sootywing (*Pholisora mejicana*)
___ Mohave Sootywing (*Hesperopsis libya*)
___ Saltbush Sootywing (*Hesperopsis alpheus*)
___ Brown-banded Skipper (*Timochares ruptifasciatus*)
___ White-patched Skipper (*Chiomara georgina*)
___ Dreamy Duskywing (*Erynnis icelus*)
___ Sleepy Duskywing (*Erynnis brizo*)
___ 'Arizona' Juvenal's Duskywing (*Erynnis juvenalis clitus*)
___ Rocky Mountain Duskywing (*Erynnis telemachus*)
___ Meridian Duskywing (*Erynnis meridianus*)
___ Scudder's Duskywing (*Erynnis scudderi*)
___ Horace's Duskywing (*Erynnis horatius*)
___ Mournful Duskywing (*Erynnis tristis tatius*)
___ Mottled Duskywing (*Erynnis martialis*)
___ Pacuvius Duskywing (*Erynnis pacuvius*)
___ Funereal Duskywing (*Erynnis funeralis*)
___ Afranius Duskywing (*Erynnis afranius*)
___ Persius Duskywing (*Erynnis persius*)
___ Texas Powdered-Skipper (*Systasea pulverulenta*)
___ Arizona Powdered-Skipper (*Systasea zampa*)
___ Common Streaky-Skipper (*Celotes nessus*)
___ Grizzled Skipper (*Pyrgus centaureae*)
___ Mountain Checkered-Skipper (*Pyrgus xanthus*)
___ Small Checkered-Skipper (*Pyrgus scriptura*)
___ Common Checkered-Skipper (*Pyrgus communis*)
___ White Checkered-Skipper (*Pyrgus albescens*)
___ Desert Checkered-Skipper (*Pyrgus philetas*)
___ Tropical Checkered-Skipper (*Pyrgus oileus*)
___ Erichson's White-Skipper (*Heliopyrgus domicella*)
___ Northern White-Skipper (*Heliopetes ericetorum*)

Skipperlings (Subfamily Heteropterinae)

___ Russet Skipperling (*Piruna pirus*)
___ Four-spotted Skipperling (*Piruna polingii*)

Giant-Skippers (Subfamily Megathyminae)

___ Orange Giant-Skipper (*Agathymus neumoegeni*)
___ Poling's Giant-Skipper (*Agathymus polingi*)
___ Arizona Giant-Skipper (*Agathymus aryxna*)

___ Mary's Giant-Skipper (*Agathymus mariae*)
___ Yucca Giant-Skipper (*Megathymus yuccae*)
___ Ursine Giant-Skipper (*Megathymus ursus*)
___ Strecker's Giant-Skipper (*Megathymus streckeri*)

Grass Skippers (Subfamily Hesperiinae)
___ Least Skipper (*Ancyloxypha numitor*)
___ Tropical Least Skipper (*Ancyloxypha arene*)
___ Garita Skipperling (*Oarisma garita*)
___ Edwards' Skipperling (*Oarisma edwardsii*)
___ Orange Skipperling (*Copaeodes aurantiaca*)
___ Sunrise Skipper (*Adopaeoides prittwitzi*)
___ Brazilian Skipper (*Calpodes ethlius*)
___ Large Roadside-Skipper (*Amblyscirtes exoteria*)
___ Cassus Roadside-Skipper (*Amblyscirtes cassus*)
___ Bronze Roadside-Skipper (*Amblyscirtes aenus*)
___ Oslar's Roadside-Skipper (*Amblyscirtes oslari*)
___ Texas Roadside-Skipper (*Amblyscirtes texanae*)
___ Slaty Roadside-Skipper (*Amblyscirtes nereus*)
___ Nysa Roadside-Skipper (*Amblyscirtes nysa*)
___ Dotted Roadside-Skipper (*Amblyscirtes eos*)
___ Common Roadside-Skipper (*Amblyscirtes vialis*)
___ Orange-headed Roadside-Skipper (*Amblyscirtes phylace*)
___ Eufala Skipper (*Lerodea eufala*)
___ Clouded Skipper (*Lerema accius*)
___ Fiery Skipper (*Hylephila phyleus*)
___ Uncas Skipper (*Hesperia uncas*)
___ Juba Skipper (*Hesperia juba*)
___ Colorado Branded Skipper (*Hesperia colorado*)
___ Apache Skipper (*Hesperia woodgatei*)
___ Pahaska Skipper (*Hesperia pahaska*)
___ Green Skipper (*Hesperia viridis*)
___ Nevada Skipper (*Hesperia nevada*)
___ Rhesus Skipper (*Polites rhesus*)
___ Carus Skipper (*Polites carus*)
___ Sandhill Skipper (*Polites sabuleti*)
___ Draco Skipper (*Polites draco*)
___ Tawny-edged Skipper (*Polites themistocles*)
___ Crossline Skipper (*Polites origenes*)
___ Sachem (*Atalopedes campestris*)
___ Hobomok Skipper (*Poanes hobomok*)
___ Taxiles Skipper (*Poanes taxiles*)
___ Umber Skipper (*Poanes melane*)
___ Morrison's Skipper (*Stinga morrisoni*)
___ Woodland Skipper (*Ochlodes sylvanoides*)
___ Yuma Skipper (*Ochlodes yuma*)
___ Snow's Skipper (*Paratrytone snowi*)
___ Delaware Skipper (*Anatrytone logan*)
___ Simius Roadside-Skipper (*Notamblyscirtes simius*)
___ Dun Skipper (*Euphyes vestris*)
___ Dusted Skipper (*Atrytonopsis hianna*)
___ Deva Skipper (*Atrytonopsis deva*)
___ Moon-marked Skipper (*Atrytonopsis lunus*)
___ Viereck's Skipper (*Atrytonopsis vierecki*)
___ White-barred Skipper (*Atrytonopsis pittacus*)
___ Python Skipper (*Atrytonopsis python*)
___ Sheep Skipper (*Atrytonopsis edwardsi*)

SWALLOWTAILS (FAMILY PAPILIONIDAE)
Parnassians (Subfamily Parnassiinae)
___ Smintheus Parnassian (*Parnassius smintheus*)
Swallowtails (Subfamily Papilioninae)
___ Pipevine Swallowtail (*Battus philenor*)

Above:
BRAZILIAN SKIPPER

___ Polydamas Swallowtail (*Battus polydamas*)
___ Old World Swallowtail (*Papilio machaon*)
___ Black Swallowtail (*Papilio polyxenes*)
___ Anise Swallowtail (*Papilio zelicaon*)
___ Indra Swallowtail (*Papilio indra*)
___ Giant Swallowtail (*Papilio cresphontes*)
___ Broad-banded Swallowtail (*Papilio astyalus*)
___ Ornythion Swallowtail (*Papilio ornythion*)
___ Western Tiger Swallowtail (*Papilio rutulus*)
___ Pale Swallowtail (*Papilio eurymedon*)
___ Two-tailed Swallowtail (*Papilio multicaudata*)
___ Palamedes Swallowtail (*Papilio palamedes*)

WHITES AND SULPHURS (FAMILY PIERIDAE)
Sulphurs (Subfamily Coliadinae)
___ Lyside Sulphur (*Kricogonia lyside*)
___ Dainty Sulphur (*Nathalis iole*)
___ Barred Yellow (*Eurema daira*)
___ Boisduval's Yellow (*Eurema boisduvaliana*)
___ Mexican Yellow (*Eurema mexicana*)
___ Salome Yellow (*Eurema salome*)
___ Tailed Orange (*Pyrisitia proterpia*)
___ Little Yellow (*Pyrisitia lisa*)
___ Mimosa Yellow (*Eurema nise*)
___ Sleepy Orange (*Abaeis nicippe*)
___ Clouded Sulphur (*Colias philodice*)
___ Orange Sulphur (*Colias eurytheme*)
___ Queen Alexandra's Sulphur (*Colias alexandra*)
___ Mead's Sulphur (*Colias meadii*)
___ Scudder's Sulphur (*Colias scudderii*)
___ Southern Dogface (*Zerene cesonia*)
___ White Angled-Sulphur (*Anteos clorinde*)
___ Yellow Angled-Sulphur (*Anteos maerula*)
___ Cloudless Sulphur (*Phoebis sennae*)
___ Large Orange Sulphur (*Phoebis agarithe*)
___ Orange-barred Sulphur (*Phoebis philea*)

Whites (Subfamily Pierinae)
___ Desert Orangetip (*Anthocharis cethura*)
___ Sara Orangetip (*Anthocharis sara*)
___ Large Marble (*Euchloe ausonides*)
___ Olympia Marble (*Euchloe olympia*)
___ Desert Marble (*Euchloe lotta*)
___ Florida White (*Glutophrissa drusilla*)
___ Pine White (*Neophasia menapia*)
___ Chiricahua White (*Neophasia terlooii*)
___ Margined White (*Pieris marginalis*)
___ Cabbage White (*Pieris rapae*)
___ Becker's White (*Pontia beckerii*)
___ Checkered White (*Pontia protodice*)
___ Western White (*Pontia occidentalis*)
___ Spring White (*Pontia sisymbrii*)
___ Great Southern White (*Ascia monuste*)
___ Giant White (*Ganyra josephina*)

GOSSAMER-WINGS (FAMILY LYCAENIDAE)
Coppers (Subfamily Lycaeninae)
___ Lustrous Copper (*Lycaena cupreus*)
___ Tailed Copper (*Lycaena arota*)
___ Gray Copper (*Lycaena dione*)
___ Ruddy Copper (*Lycaena rubidus*)
___ Blue Copper (*Lycaena heteronea*)
___ Bronze Copper (*Lycaena hyllus*)
___ Purplish Copper (*Lyceana helloides*)

Hairstreaks (Subfamily Theclinae)

___ Colorado Hairstreak (*Hypaurotis crysalus*)
___ Great Purple Hairstreak (*Atlides halesus*)
___ Behr's Hairstreak (*Satyrium behrii*)
___ Sylvan Hairstreak (*Satyrium sylvinus*)
___ Coral Hairstreak (*Satyrium titus*)
___ Banded Hairstreak (*Satyrium calanus*)
___ Striped Hairstreak (*Satyrium liparops*)
___ Southern Hairstreak (*Satyrium favonius*)
___ Ilavia Hairstreak (*Satyrium ilavia*)
___ Poling's Hairstreak (*Satyrium polingi*)
___ Soapberry Hairstreak (*Phaeostrymon alcestis*)
___ Silver-banded Hairstreak (*Chlorostrymon simaethis*)
___ Bramble Hairstreak (*Callophrys affinis*)
___ Sheridan's Hairstreak (*Callophrys sheridanii*)
___ Juniper Hairstreak (*Callophrys gryneus*)
___ Xami Hairstreak (*Callophrys xami*)
___ Sandía Hairstreak (*Callophrys mcfarlandi*)
___ Thicket Hairstreak (*Callophrys spinetorum*)
___ Brown Elfin (*Callophrys augustinus*)
___ Desert Elfin (*Callophrys fotis*)
___ Hoary Elfin (*Callophrys polios*)
___ Henry's Elfin (*Callophrys henrici*)
___ Western Pine Elfin (*Callophrys eryphon*)
___ Gray Hairstreak (*Strymon melinus*)
___ Leda Ministreak (*Ministrymon leda*)
___ Arizona Hairstreak (*Erora quaderna*)

Blues (Subfamily Polyommatinae)

___ Cassius Blue (*Leptotes cassius*)
___ Marine Blue (*Leptotes marina*)
___ Western Pygmy-Blue (*Brephidium exilis*)
___ Cyna Blue (*Zizula cyna*)
___ Eastern Tailed-Blue (*Cupido comyntas*)
___ Western Tailed-Blue (*Cupido amyntula*)
___ Spring Azure (*Celastrina ladon*)
___ Ceraunus Blue (*Hemiargus ceraunus*)
___ Square-spotted Blue (*Euphilotes battoides*)
___ Ellis' Blue (*Euphilotes ellisii*)
___ Rocky Mountain Dotted-Blue (*Euphilotes ancilla*)
___ Rita Blue (*Euphilotes rita*)
___ Spalding's Blue (*Euphilotes spaldingi*)
___ Arrowhead Blue (*Glaucopsyche piasus*)
___ Silvery Blue (*Glaucopsyche lygdamus*)
___ Reakirt's Blue (*Echinargus isola*)
___ Melissa Blue (*Plebejus melissa*)
___ Greenish Blue (*Plebejus saepiolus*)
___ Boisduval's Blue (*Plebejus icarioides*)
___ Acmon Blue (*Plebejus acmon*)
___ Arctic Blue (*Plebejus glandon*)

METALMARKS (FAMILY RIODINIDAE)

___ Fatal Metalmark (*Calephelis nemesis*)
___ Arizona Metalmark (*Calephelis arizonensis*)
___ Zela Metalmark (*Emesis zela*)
___ Ares Metalmark (*Emesis ares*)
___ Mormon Metalmark (*Apodemia mormo*)
___ Mexican Metalmark (*Apodemia mejicanus*)
___ Dury's Metalmark (*Apodemia duryi*)
___ Palmer's Metalmark (*Apodemia palmerii*)
___ Nais Metalmark (*Apodemia nais*)

ZILPA LONGTAIL

BRUSHFOOT BUTTERFLIES (FAMILY NYMPHALIDAE)

Snouts (Subfamily Libytheinae)

___ American Snout (*Libytheana carinenta*)

Monarchs (Subfamily Danainae)

___ Monarch (*Danaus plexippus*)
___ Queen (*Danaus gilippus*)
___ Soldier (*Danaus eresimus*)

Admirals and Relatives (Subfamily Limenitidinae)

___ Red-spotted Admiral (*Limenitis arthemis*)
___ Weidemeyer's Admiral (*Limenitis weidemeyerii*)
___ Viceroy (*Limenitis archippus*)
___ Arizona Sister (*Adelpha eulalia*)

Heliconians and Fritillaries (Subfamily Heliconiinae)

___ Mexican Silverspot (*Dione moneta*)
___ Gulf Fritillary (*Agraulis vanillae*)
___ Julia Heliconian (*Dryas iulia*)
___ Isabella's Heliconian (*Eueides isabella*)
___ Zebra Heliconian (*Heliconius charithonia*)
___ Variegated Fritillary (*Euptoieta claudia*)
___ Mexican Fritillary (*Euptoieta hegesia*)
___ Silver-bordered Fritillary (*Boloria selene*)
___ Freija Fritillary (*Boloria freija*)
___ Arctic Fritillary (*Boloria chariclea*)
___ Great Spangled Fritillary (*Speyeria cybele*)
___ Aphrodite Fritillary (*Speyeria aphrodite*)
___ Nokomis Fritillary (*Speyeria nokomis*)
___ Edwards' Fritillary (*Speyeria edwardsii*)
___ Northwestern Fritillary (*Speyeria hesperis*)
___ Mormon Fritillary (*Speyeria mormonia*)

Emperors (Subfamily Apaturinae)

___ Hackberry Emperor (*Asterocampa celtis*)
___ Empress Leilia (*Asterocampa leilia*)
___ Tawny Emperor (*Asterocampa clyton*)
___ Silver Emperor (*Doxocopa laure*)

Tropical Brushfoots (Subfamily Biblidinae)

___ Common Mestra (*Mestra amymone*)
___ Dingy Purplewing (*Eunica monima*)
___ Blackened Bluewing (*Myscelia cyananthe*)
___ Gray Cracker (*Hamadryas februa*)
___ Ruddy Daggerwing (*Marpesia petreus*)

True Brushfoots (Subfamily Nymphalinae)

___ American Lady (*Vanessa virginiensis*)
___ Painted Lady (*Vanessa cardui*)
___ West Coast Lady (*Vanessa annabella*)

SOLDIER

___ Red Admiral (*Vanessa atalanta*)
___ Milbert's Tortoiseshell (*Aglais milberti*)
___ California Tortoiseshell (*Nymphalis californica*)
___ Mourning Cloak (*Nymphalis antiopa*)
___ Question Mark (*Polygonia interrogationis*)
___ Satyr Comma (*Polygonia satyrus*)
___ Hoary Comma (*Polygonia gracilis*)
___ Green Comma (*Polygonia faunus*)
___ White Peacock (*Anartia jatrophae*)
___ Malachite (*Siproeta stelenes*)
___ Rusty-tipped Page (*Siproeta epaphus*)
___ Common Buckeye (*Junonia coenia*)
___ Tropical Buckeye (*Junonia genoveva nigrosuffusa*)
___ Anicia Checkerspot (*Euphydryas anicia*)
___ Dotted Checkerspot (*Poladryas minuta*)
___ Arachne Checkerspot (*Poladryas arachne*)
___ Crimson Patch (*Chlosyne janais*)
___ Definite Patch (*Chlosyne definita*)
___ Theona Checkerspot (*Chlosyne theona*)
___ Fulvia Checkerspot (*Chlosyne fulvia*)
___ Silvery Checkerspot (*Chlosyne nycteis*)
___ Gorgone Checkerspot (*Chlosyne gorgone*)
___ Bordered Patch (*Chlosyne lacinia*)
___ Sagebrush Checkerspot (*Chlosyne acastus*)
___ Tiny Checkerspot (*Dymasia dymas*)
___ Elada Checkerspot (*Texola elada*)
___ Pale-banded Crescent (*Anthanassa tulcis*)
___ Texan Crescent (*Anthanassa texana*)
___ Graphic Crescent (*Phyciodes graphica*)
___ Painted Crescent (*Phyciodes picta*)
___ Mylitta Crescent (*Phyciodes mylitta*)
___ Phaon Crescent (*Phyciodes phaon*)
___ Pearl Crescent (*Phyciodes tharos*)
___ Northern Crescent (*Phyciodes cocyta*)
___ Tawny Crescent (*Phyciodes batesii*)
___ Field Crescent (*Phyciodes pulchella*)

Leafwings (Subfamily Charaxinae)
___ Tropical Leafwing (*Anaea aidea*)
___ Goatweed Leafwing (*Anaea andria*)

Satyrs (Subfamily Satyrinae)
___ Common Ringlet (*Coenonympha tullia*)
___ Nabokov's Satyr (*Cyllopsis pyracmon*)
___ Canyonland Satyr (*Cyllopsis pertepida*)
___ Red Satyr (*Megisto rubricata*)
___ Common Wood-Nymph (*Cercyonis pegala*)
___ Mead's Wood-Nymph (*Cercyonis meadii*)
___ Great Basin Wood-Nymph (*Cercyonis sthenele*)
___ Small Wood-Nymph (*Cercyonis oetus*)
___ Red-bordered Satyr (*Gyrocheilus patrobas*)
___ Magdalena Alpine (*Erebia magdalena*)
___ Common Alpine (*Erebia epipsodea*)
___ Riding's Satyr (*Neominois ridingsii*)
___ Polixenes Arctic (*Oeneis polixenes*)
___ Melissa Arctic (*Oeneis melissa*)
___ Chryxus Arctic (*Oeneis chryxus*)
___ Alberta Arctic (*Oeneis alberta*)
___ Uhler's Arctic (*Oeneis uhleri*)

Butterfly Glossary

abdomen: the most posterior division of an insect's body

adult: the fourth life stage of Lepidoptera, during which reproduction occurs; also imago

antennae: a pair of sensory organs on the head; clubbed in butterflies

biogeography: the study of where and why organisms occur

brood: even-aged offspring of the females of a given species

butterflying: observing butterflies

caterpillar: the second Lepidoptera life stage, during which growth occurs; see larva

chrysalis: the skin within which metamorphosis occurs during the third life stage of Lepidoptera; aka chrysalid, pupa

community: an assemblage of organisms that interact with each other in a particular area

corridor: a route that allows dispersal of butterflies from one place to another

diapause: a period of dormancy often triggered by seasonally cold or dry weather

diversity: the number of different species present in some defined area

dormancy: a period of inactivity

ecoregion: a geographic region with internally consistent landforms, plants, animals, and climate

ecology: the study of factors that influence abundance and distribution of organisms

ecosystem: a biological community in relation to its physical environment

egg: the first life stage of Lepidoptera; also ovum

exoskeleton: an outer skeleton, as in insects, compared to humans' endo- (or inner) skeleton

generation: a discrete, complete life cycle from adult to egg, larva, pupa, and adult again

habitat: the place where an organism normally lives

head: the most anterior major body part of an insect

herbivore: an animal that eats only plants

hibernation: a dormancy period through a winter season; see diapause, overwinter

hilltopping: a behavior in which males seek females on hilltops

host plant: the plant preferred by caterpillars of a butterfly species, where females place eggs; aka food plant

immature: any stage prior to the adult

instar: a developmental step within the larval stage that results in shedding of the exoskeleton and replacement with a larger one

larva: the second Lepidoptera life stage, during which growth occurs; see caterpillar

Lepidoptera: the insect order containing butterflies and moths

lepidopterist: anyone interested in butterflies or moths

metamorphosis: a change in form, as when a caterpillar changes to an adult butterfly

migrate: to change location by directed dispersal, usually in large numbers

mutualistic: describing a biological interaction between individuals of two different species, in which each individual derives a benefit

myrmecophily: literally "ant-loving," as in most blues, coppers, and hairstreaks

nectaring: a behavior in which butterflies siphon nectar from flowers

overwinter: to survive winter in-place; to hibernate

oviposit: to place an egg

parasite: an organism that lives off another organism without killing it

patrolling: a mate location strategy in which males fly actively to seek females

perching: a mate location strategy in which males perch and wait for females to pass

population: a group of interbreeding individuals of the same species, separated in space or time from other groups of the same species

predator: an animal that eats animals; a carnivore

prey: an animal that typically is eaten by other animals

proboscis: the coiled, straw-like tube used by adult butterflies to siphon nectar or moisture

pupa: the third life stage of Lepidoptera, during which metamorphosis occurs; called a chrysalis in butterflies

race: within a species, a distinct population that may be geographically limited; see subspecies

refuge: an area that remains unchanged while nearby areas have changed, allowing survival of organisms that no longer remain in surrounding areas

scale: a flattened, shingle-like hair attached to a wing, providing color and pattern to wing surfaces

species: a group of interbreeding organisms that are different from other such reproductive groups

stray: an individual far outside of its normal range

subspecies: a distinct subset of a species; may be geographic in origin; see race

tail: an elongated protrusion from the rear part of the hindwing, present in only a few species

territory: living space that enhances reproduction or survival of an individual

thorax: the middle body division of insects

vagrant: another term for stray

warning colors: usually black and red, advertising a creature's distastefulness and thereby discouraging predation

Above:
SALOME YELLOW

Recommended Reading

REGIONAL BUTTERFLY BOOKS

Bailowitz, R. A., and J. P. Brock. 1991. *Butterflies of Southeastern Arizona*. Sonoran Arthropod Studies, Inc. Tucson, AZ.

Bailowitz, R. A., and D. Danforth. 1997. *70 Common Butterflies of the Southwest*. Southwest Parks & Monuments Association. Tucson, AZ.

Brock, J. P. 2008. *Butterflies of the Southwest*. Río Nuevo Publishers. Tucson, AZ.

Cary, S. J., and R. Holland. 1992. "New Mexico Butterflies: Checklist, Distribution and Conservation." *Journal of Research on the Lepidoptera* 31(1): 57-82.

Ferris, C. D., and F. M. Brown. 1981. *Butterflies of the Rocky Mountain States*. University of Oklahoma Press. Norman, OK.

Stewart, B., P. Brodkin, and H. Brodkin. 2001. *Butterflies of Arizona: A Photographic Guide*. West Coast Lady Press. Arcata, CA.

Toliver, M. E., R. Holland, and S. J. Cary. 2003. *Distribution of Butterflies in New Mexico (Lepidoptera: Hesperioidea and Papilionoidea)*. 3rd ed. Published by R. Holland. Albuquerque, NM.

OTHER RELEVANT BOOKS

Carter, J. L. 1997. *Trees and Shrubs of New Mexico*. Mimbres Publishing. Silver City, NM.

Greene, P. 2006. *Mike Butterfield's Guide to the Mountains of New Mexico*. New Mexico Magazine. Santa Fe, NM.

Ivey, R. D. 2003. *Flowering Plants of New Mexico*. 4th ed. Published by R. D. Ivey. Albuquerque, NM.

Scott, J. A. 1986. *The Butterflies of North America: A Natural History and Field Guide*. Stanford University Press. Stanford, CA.

SELECTED FIELD GUIDES

Brock, J. P., and K. Kaufman. 2006. *Kaufman Field Guide to Butterflies of North America*. Houghton Mifflin. Boston, MA.

Glassberg, J. 2001. *Butterflies Through Binoculars: The West*. Oxford University Press. New York, NY.

Parmeter, J., B. Neville, and D. Emkalns. 2002. *New Mexico Bird Finding Guide*. New Mexico Ornithological Society. Albuquerque, NM.

Zimmerman, D. A., and C. M. Hunter. 2006. *Conspicuous Butterflies of the Gila National Forest, New Mexico and Surrounding Areas*. Gila Pocket Guides. Biota Press. Silver City, NM.

METHODS & TOOLS

Lepidopterists' Society. 1996. "Statement on Collecting." http://www.lepsoc.org/statement_on_collecting.php.

Winter, W. D., Jr. 2000. *Basic Techniques for Observing and Studying Moths and Butterflies.* Memoir No. 5. The Lepidopterists' Society.

NAMES

North American Butterfly Association. *English Names of North American Butterflies Occurring North of Mexico.* http://www.naba.org/pubs/enames.html.

Pelham, J. P. 2008. "A Catalogue of the Butterflies of the United States and Canada." *Journal of Research on the Lepidoptera 40.*

GARDENING

Tekulsky, M. 1985. *The Butterfly Garden.* The Harvard Common Press. Boston, MA.

Xerces Society and Smithsonian Institution. 1998. *Butterfly Gardening, Creating Summer Magic in Your Garden.* Sierra Club Books. San Francisco, CA.

BIODIVERSITY

Opler, P. A. 1995. "Species Richness and Trends of Western Butterflies and Moths." 172-174 in LaRoe, E. T., G. S. Farris, C. E. Puckett, P. D. Doran, and M. J. Mac. *Our Living Resources: A Report to the Nation on Distribution, Abundance, and Health of U.S. Plants, Animals, and Ecosystems.* U.S. Department of the Interior, National Biological Service. Washington, D.C.

MAPS

New Mexico Road & Recreation Atlas, Benchmark Maps, Bedford, OR.

New Mexico State Highway Map, New Mexico Department of Transportation, Santa Fe, NM.

Above:
BROWN-BANDED SKIPPER
Opposite page, below:
CASSIUS BLUE

New Mexico has 35 wonderful state parks where visitors can enjoy butterflies and other outdoor recreational pursuits. Information about these parks can be found at www.nmparks.com. Contact individual parks at the telephone numbers below.

Bluewater Lake: 505-876-2391
Bottomless Lakes: 575-624-6058
Brantley Lake: 575-457-2384
Caballo Lake: 575-743-3942
Cerrillos Hills: 505-474-0196
Cimarron Canyon: 575-377-6271
City of Rocks: 575-536-2800
Clayton Lake: 575-374-8808
Conchas Lake: 575-868-2270
Coyote Creek: 575-387-2328
Eagle Nest Lake: 575-377-1594
Elephant Butte Lake: 575-744-5923
El Vado Lake: 575-588-7247
Fenton Lake: 575-829-3630
Heron Lake: 575-588-7240
Hyde Memorial: 505-983-7175
Leasburg Dam: 575-524-4068
Living Desert Zoo and Gardens: 575-887-5516
Manzano Mountains: 505-847-2820
Mesilla Valley Bosque: 575-523-4398
Morphy Lake: 575-387-2328
Navajo Lake: 505-632-2278
Oasis: 575-356-5331
Oliver Lee Memorial: 575-437-8284
Pancho Villa: 575-531-2711
Percha Dam: 575-743-3942
Río Grande Nature Center: 505-344-7240
Rockhound: 575-546-6182
Santa Rosa Lake: 575-472-3110
Storrie Lake: 505-425-7278
Sugarite Canyon: 575-445-5607
Sumner Lake: 575-355-2541
Ute Lake: 575-487-2284
Vietnam Veterans Memorial: 575-377-2293
Villanueva: 575-421-2957

State Map

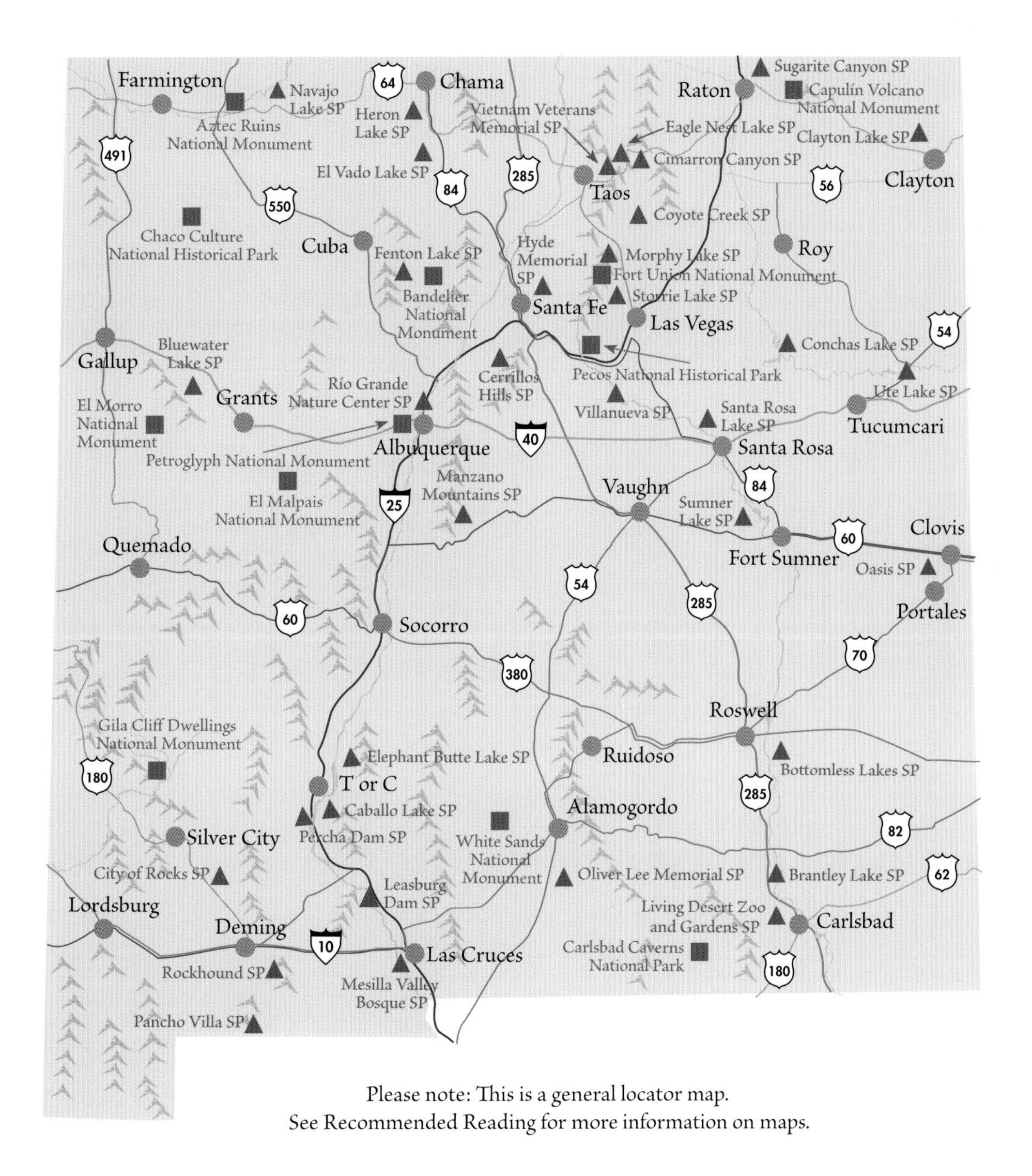

Please note: This is a general locator map.
See Recommended Reading for more information on maps.

Index

Numbers in italics refer to pages with photos or illustrations and sometimes text. Names in bold refer to butterflies.

A Mountain, 131
Abert, James, 93
acacia, 113
Acacia Skipper, *113*
acanthus, 125, 131
Acmon Blue, *32*
Afranius Duskywing, *42*
agaves, 115, 118, 122, 133
Aguirre Springs Recreation Area, 121–22
Alamogordo, 31, 134
Alberta Arctic, *73*, 74–75
Albuquerque, 33, 59
alfalfa, 18
Anicia Checkerspot:
  'Alpine' Anicia Checkerspot, *83*
  'Chuska Mountains' Anicia Checkerspot, 98
  'Front Range' Anicia Checkerspot, 78, *79*
  'Hermosa' Anicia Checkerspot, *110*
  'Sacramento Mountains' Anicia Checkerspot, *136*
amaranths, 36
American Lady, *21*
American Snout, 22, *23*
Animas Valley, 111
Anise Swallowtail, 102, *103*
antelope sage, 42
ants, 7, 91–92, 105
Apache Box, 109–10
Apache Creek, 109
Apache Junction, 103
Apache Skipper, 86, *109*
Aphrodite Fritillary, *78*
Arachne Checkerspot, *42*
Arctic Blue, *89*
Arctic Fritillary, *81*
Arctic Zone, 26–27, 29, 77, 79
Ares Metalmark, *112*
Arizona, 109–11, 115, 117
Arizona Giant-Skipper, *115*
Arizona Hairstreak, *109*
'Arizona' Juvenal's Duskywing, *111*
Arizona Metalmark, *114*
'Arizona' Mournful Duskywing, *123*
Arizona Powdered-Skipper, *124*
'Arizona' Red-Spotted Admiral, *127*
Arizona Sister, *41*
Arizona Skipper, *114*
Arrowhead Blue, *98*, 99
Artesia, 18
aspen, 53
asters, 14, 40, 53, 97, 107
Aztec, 101

BARRED YELLOW

Banded Hairstreak, 90, *91*
Bandelier National Monument, 47, 84
Barred Yellow, *152*
Bear Trap Campground, 57
bearberry, 84
beardtongues, 42, 70
Bearwallow Mountain, 104
Becker's White, *95*
beebalm, 14, 50
Behr's Hairstreak, *99*
Big Arsenic Springs, 97
Big Tesuque Campground, 79
biodiversity, 63
Bitter Lake National Wildlife Refuge, 65
Black Swallowtail, 34, *35*, 93
Blackened Bluewing, *12*
Blue Copper, *90*
blue grama grass, 38, 104
Boisduval's Blue, *55*, 104
Boisduval's Yellow, *130*
Bordered Patch, 80, *122*
Bottomless Lakes State Park, 36, 65
Bramble Hairstreak, *47*
Brazilian Skipper, *141*
breeding. *See* mating habits
Broad-banded Swallowtail, *157*
Bronze Copper, *72*
Bronze Roadside-Skipper, *30*, 31
Brown-banded Skipper, *149*
Brown Elfin, *107*
Brown, F. Martin, 75
Brown Longtail, *118*
buckwheat brush, 101
bunchgrass, 85
Bureau of Land Management (BLM), 87, 97, 121–22, 139
Burnt Mesa, 47, 84
Burro Mountains, 109
butterflies: anthropology and, 96; bodies of, *8*, 9–10; camouflage and, 9, 105; colors of, 9, 105; crucial role of, 7; naming of, 7; proboscis and, 10, 14, 37, 50; senses of, 10; vision of, 10; wings and, 8–9, 37, 41, 105
butterfly bush, 14

Cabbage White, *18*
Caicus Skipper, *113*

JULIA HELICONIAN

California Tortoiseshell, *47*
Camp May, 85
Canadian River, 69–70, 93
Canadian Zone, 26–27, 29, 52–61, 73, 77, 95, 102
Canyonland Satyr, *48*
Capitán Mountains, 53, 130, 133, 135
Caprock Escarpment, 68–69
Caprock Park, 38, 68
Carlsbad, 121, 129–30, 132
Carlsbad Caverns National Park, 132–33
Carpenter, William, 92–93
Carson National Forest, 88
Carus Skipper, *124*
Cassius Blue, *148*
Cassus Roadside-Skipper, *46*
caterpillars, 10, *11*, 14, 17, 24–25, 37–39, 41, 78, 92, 105, 112–13, 126. *See also* specific butterflies
Ceraunus Blue, *114*
Cerrillos Hills State Park, 32
Cerro, 97
Chaco Culture National Historical Park, 100
Chama, 39, 90, 101
Chama River. *See* Río Chama
Checkered White, *19*, 25
Cherry Creek Campground, 107
Chicorica Creek, 77
Chihuahuan Plateau, *62*, 63, 107, 121, 130
Chiricahua Mountains, 115
Chiricahua White, *115*
chokecherry, 33, 73, 90–91
chrysalids, 10, *11*, 67, 114, 126
Chryxus Arctic, *85*
Chuska Mountains, 95, 98, 103
Cibola National Forest, 70, 99
Cimarron Canyon State Park, 78
cinquefoils, 54
Clanton Draw, 111, 115
Clayton, 72–73, 77
Clayton Lake State Park, 72
Cliff, 125
cliffrose, 100
climate, 22–24, *27*, 31, 37–39, 43, 61, 74, 95, 97, 110, 114, 121
Cloudcroft, 134, 136
Clouded Skipper, *129*
Clouded Sulphur, *18*
Cloudless Sulphur, *24*
Clover Blue, 104
Cloverdale Creek, 111

Above:
POLYDAMAS SWALLOWTAIL

clovers, 83, 88
Clovis, 66, 68
Cockerell, Theodore, *80*, 81
Colfax County, 74
Colin Neblett Wildlife Area, 78
collecting specimens, 12–13, 29, 49, 86, 93, 108, 118, 134, 139
Colorado, 73, 75, 78, 80, 103, 108
Colorado Branded Skipper, 38, *76*, *77*, 104
Colorado Hairstreak, *41*
Colorado Plateau, *62*, 63, *77*, 95, 101
Colorado River, 63
Common Alpine, *88*
Common Buckeye, 22, *23*
Common Checkered-Skipper, *20*, 25, 114
Common Mestra, *64*, 65
common reed, 97
Common Ringlet, *77*
Common Roadside-Skipper, *86*, 87
Common Sootywing, *36*, *77*
Common Streaky-Skipper, *124*
Common Wood-Nymph, *71*
Conchas Lake State Park, 31, 69
coneflowers, 40, 97
Cooke's Range, 122
Coral Hairstreak, *91*
Cottonwood Canyon, 111
cottonwoods, 132
Coyote Creek State Park, 78
Crimson Patch, *158*
Crossline Skipper, *74*
cushion buckwheat, 101
cutleaf coneflower, 57, 78, 90, 137
Cyna Blue, *129*
Dainty Sulphur, *18*, 19, 25
daisy family, 14, 19, 53, 107
dandelions, 90, 135
Dátil Range, 40
Definite Patch, *131*
Delaware Skipper, *67*
Deming, 121, 124
Desert Borderlands, *61*, 120–37
Desert Checkered-Skipper, *114*
Desert Cloudywing, *116*
Desert Elfin, *100*
desert hackberry, 111
Desert Marble, *107*
Desert Orangetip, *110*
Desert Viceroy, *125*
Deva Skipper, *107*
diapause, 34, 126
Dingy Purplewing, *162*
disturbance, 43, *44*, *45*
diversity, 28
docks, 72, 90
Dog Canyon, 127
dogbane, 14, 70
Dorantes Longtail, *118*
dormancy, 121, 126
Dotted Checkerspot, *70*, 93
Dotted Roadside-Skipper, *124*
doveweed, 71
Draco Skipper, *89*
Dreamy Duskywing, *52*, 53
Dripping Springs, 122, 131
drought, 121, 126, 139
Drusius Cloudywing, *5*

Dull Firetip, *112*
Dun Skipper, 50, *51*
Dury's Metalmark, *131*, 132
Dusted Skipper, *84*

Eagle Nest, 39, 78
Eagle Nest Lake State Park, 54
Eastern Plains, *61*, 63–74
Eastern Tailed-Blue, *125*
ecoregions, 61–63, 65, 77, 95, 103, 107, 130, 136
ecosystems, 7, 14, 25, 37, 42–45, 61, 92, 103, 119
Edwards' Fritillary, *78*
Edwards' Skipperling, *48*
Edwards, William E., 93, 108
egg laying, 10, *11*, 58–59, 107, 110, 129
El Paso, Texas, 134
El Vado Lake State Park, 40
Elada Checkerspot, *125*
Elena Gallegos recreation area, 35
Ellis' Blue, *100*, 101
Emory Pass, 107, 109
Emory, William, 93
Empress Leilia, *111*
entomologists, 49, 80, 86, 108, 134. *See also* individual names
Erichson's White-Skipper, *116*
Eufala Skipper, 130, *131*
Experiment Station Bulletin (NMSU), 29

Farmington, 95
Fatal Metalmark, *127*
federal agencies, 139
feeding habits, 10, 14, 17–21, 24–25, 31–33, 37–38, 41–42, 47, 60, 101, 112, 119. *See also* specific butterflies
Fendler's buckbrush, 47, 85
Fenton Lake State Park, *45*, 84, 87
field bindweed, 20
Field Crescent, *40*
field guides, 6–7, 12–13
Fiery Skipper, *129*
fires, 45, 47, 84, 103
Fishers Peak, 73
Florida White, *165*
flying habits, 9–10, 37, 41, 47, 60, 69, 78, 126. *See also* specific butterflies
forget-me-nots, 83
Fort Wingate, 99
Four-spotted Skipperling, 136, *137*
four-wing saltbush, 36
Freija Fritillary, *88*
Fulvia Checkerspot, *11*, *32*
Funereal Duskywing, *22*

Gain Access Into Nature (GAIN), 139
Gallinas Canyon, 49–50
Gallinas Mountains, 34
Gambel oaks, 39, *41*, 45, 73, 90
gardens, 14, *15*, 18, 33
Garita Skipperling, 38, *54*
Geronimo Pass, 111
Giant Swallowtail, 128, *129*
Giant White, *6*
Gila country, 53, 55, 63, 107
Gila National Forest, 103–04, 107, 109
Gila River, 29, 125
Goatweed Leafwing, *71*
Gold Hill, 81

Golden-banded Skipper, *106*, 107
Golden-headed Scallopwing, *124*
gooseberries, 97
goosefoots, 36
Gorgone Checkerspot, *66*, 67
Grant County, 107, 109
Grants, 100–101
Graphic Crescent, *131*
grasses, 31, *38*, 42, 45, 54, 74, 85, 97, 103, 113. *See also* specific names
Grasshopper Spring, 99
Gray Copper, *72*
Gray Cracker, frontispiece
Gray Hairstreak, *17*
Great Basin, 95, 99, 137
Great Basin Wood-Nymph, *98*
Great Plains, 61–63, 65, 72, 77, 121, 130, 136
Great Purple Hairstreak, *123*
Great Southern White, *166*
Great Spangled Fritillary, *92*
Green Comma, *78*, 79
Green Skipper, *31*, 38
Greenish Blues, 88, *89*
Grizzled Skipper, *81*
Guadalupe Canyon, 117–18
Guadalupe Mountain National Park, 133
Guadalupe Mountains, 130–31, 133
Guadalupe Peak, 130, 133
Gulf Fritillary, *67*
habitats, 26, 37, 43–45, 79, 95, 103, 112, 121; arroyos, 32, 113–14, 123–25, 131–32; canyons, 115–16, 119, 127–28, 131; desert, 63, 121, 125, 128, 132–33; disturbed, 19–20, 25, 36, 42–45, 129; dry, 40, 63; fields, 98; grasslands, 31, 38, 42, 53, 63, 71, 73, 77, 85, 95, 98, 101, 121–22; hilltops, 32, 38, 53, 60, 109, 122–23, 133; lowlands, 69, 124; meadows, 53–55, 72, 77, 81, 83, 85, 88, 90, 135–36; mountains, 28, 31, 135, 137; riparian, 32, 48, 50, 53, 55, 57–58, 60, 63, 65, 67, 69, 71, 83, 87, 90, 109, 111, 125–26, 130; rocky, 31–32, 124; saline, 36; savannas, 39, 42, 46, 48, 84, 99, 101–02, 104, 107; woodlands, 45, 53, 73, 78, 87–88, 102, 107, 121, 135

BROAD-BANDED SWALLOWTAIL

Hackberry Emperor, *70*
Hadley Draw, 122
Hamilton Mesa, 79
Hammock Skipper, *129*
Henry's Elfin, *131*, 132
Henshaw, Henry, 93
hibernation, 9, 37, 50, 71, 79
Highlands University, 80
Hillsboro, 33, 109
Hoary Comma, *53*
Hoary Elfin, *84*
Hobbs, 129
Hobomok Skipper, *73*, 74, 136
Holy Ghost Campground, 79
Holy Ghost Canyon, 53
honey mesquite, 121
Hopewell Lake, 88
Hopi Indian culture, 96
hops, 50
hoptree, 33
Horace's Duskywing, *68*
host plants, *7*, *11*, 14, 17–21, 25, 37, 42–43, 55, 60, 101, 102, 133. *See also* specific names
Howard, Winslow, 108
Hudsonian Zone, 26–27, 29, 77, 81
Humphries Wildlife Management Area, 90–91, 101
Hyde, Benjamin and Helen, *75*
Hyde Memorial State Park, 75

Ilavia Hairstreak, *109*
Indra Swallowtail, 103
insects, 8–9, 29, 60, 75, 80, 86, 105, 119, 126, 134
iris, 84
Isabella's Heliconian, title page

Jackson Lake, 95
Jémez Mountains, 53, 83–86, 109
Juba Skipper, *95*
Julia Heliconian, *153*
Juniper Hairstreak, *32*, 93
juniper, 32, 53, 104, 107

katsinas, 96
Kingston, 109
Kiowa National Grassland, 70
Knaus, Warren, 134

La Cueva recreation area, 35
Large Marble, *88*
Large Orange Sulphur, *128*
Large Roadside-Skipper, *113*
larvae, 10, *11*, 25, 34, 37, 40, 43, 47, 60, 92, 105. *See also* caterpillars and feeding habits
Las Cruces, 29, 121, 129–31
Las Vegas, 31, 39, 50, 78, 80
Last Chance Canyon, 131

Above:
CRIMSON PATCH

Least Skipper, *72*
Leda Ministreak, *121*
legumes, 18–19, 24, 42, 45, 55, 59, 89
lepidopterists, 75, 108. *See also* individual names
life cycles, 10–11, 14, 24–25, 34, 79, 105, 126
life zones, 26–27, 29, 53, 55, 61, 77, 95, 136
Lincoln National Forest, 40, 136
little bluestem grass, 84
Little Horse Mesa, 73, 77
Little Yellow, *66*
Llano Estacado, 17, 68
locust, 45, 50
Long-tailed Skipper, *118*
Los Alamos, 85
Lower Gallinas Campground, 107
Lower Sonoran Zone, 26
lupines, 45, 55, 99
Lustrous Copper, *82*
Lyside Sulphur, *128*

Madre Mountain, 40
Magdalena Alpine, *82*
Magdalena Mountains, 46, 49
Malachite, *138*
manzanita, 107
Manzano Mountains, 39
Margined White, *58*
Marine Blue, *19*
marmots, 83
Mary's Giant-Skipper, *133*
mat-plant, 65
mating habits, 9–10, *11*, 22, 25, 37, 46, 57, 60, 110, 132. *See also* specific butterflies
Maxwell National Wildlife Refuge, 72
McMillan Campground, 107
Mead, Theodore, 93
Mead's Sulphur, *82*, 83
Mead's Wood-Nymph, 38, *104*
Melissa Blue, *19*
Meridian Duskywing, *123*
Merriam, C. Hart, 29
Mesa Verde National Park, 103
Mesilla, 29
metamorphosis, 10
Mexican Cloudywing, *59*
Mexican Fritillary, title page
Mexican Metalmark, *40*
Mexican Silverspot, *12*
Mexican Sootywing, *77*
Mexican Yellow, *24*
Mexico, 7, 23–24, 38, 63, 93, 109, 115, 117, 128, 137
migration, 23
Milbert's Tortoiseshell, *58*
milkweeds, 14, 38, 78, 132
Mills Canyon, 70
Mimosa Yellow, *130*
Mississippi River, 63, 69
mistletoes, 40, 123
'Mogollón' Common Ringlet, *103*
Mogollón Creek, 125
Mogollón Highlands, *62*, 63, 103–04, 107
Mogollón Rim, 109
Mohave Sootywing, *94*, 95
Monarch, *22*, *23*, 132
monsoon season, 110, 114, 117, 121, 132
Moon-marked Skipper, *113*

Mormon Fritillary, *81*
Mormon Metalmark, *101*
Morrison's Skipper, *46*
moths, 9, 41, 49, 105, 108, 126
Mottled Arctic, *83*
Mottled Duskywing, *74*
Mount Sedgwick, 46, 102
Mount Taylor, 54, 101–02
Mountain Checkered-Skipper, *54*
mountain mahogany, 99
Mourning Cloak, *21*
mustards, 34, 58, 68, 81, 95, 107, 110
Mylitta Crescent, 50, *51*
myrmecophily, *92*

Nabokov, Vladimir, 48
Nabokov's Satyr, *118*
Nais Metalmark, *47*
National Park Service, 139
Native Americans, 96, 139
native plants, 14, *15*
naturalists, 29, 75, 86, 93, 108. *See also* individual names
Navajo Dam, 95
Navajo Nation, 98
nectar sources, 14, 31–32, 37–38, 47, 97, 136. *See also* specific plants
Ned Houk Park, 66
Negrito Creek, 48
netleaf hackberry, 70, 125
Nevada Skipper, *89*
New Mexico buckeye, 131
New Mexico Department of Game and Fish, 78, 87, 90, 95, 139
New Mexico Native Plant Society, 15
New Mexico Ornithological Society, 117
New Mexico State Land Office, 139, 150
New Mexico State Parks, 78, 139, 150
New Mexico State University (NMSU), 29, 80
New Mexico Territory, 93
Nokomis Fritillary, *48*, 80
North American Butterfly Association (NABA), *7*
North-Central Mountains, *61*, *73*, 76–92
Northern Cloudywing, *42*
Northern Crescent, *52*, 53
Northern Giant Flag Moth, 108
Northern White-Skipper, *94*, 95
Northwest Plateau, *61*, 94–104
Northwestern Fritillary, 104
'Capitán Mountains' Northwestern Fritillary, *136*
'Front Range' Northwestern Fritillary, *78*
'Jémez' Northwestern Fritillary, *85*, 86
'Raton Mesa' Northwestern Fritillary, *74*
Nysa Roadside-Skipper, *64*, 65

oaks, 68, 70, 107, 109, 111–12, 123, 133. *See also* Gambel oaks
Oasis State Park, 67
observing, 12–13, 17, 19, 26, 139
offspring, 17, 36, 53, 123
O'Keeffe, Georgia, 87
Old World Swallowtail, 93, *102*
Oliver Lee Memorial State Park, 31, 127
Olympia Marble, *68*
Orange-barred Sulphur, *130*
Orange Giant-Skipper, *122*
Orange-headed Roadside-Skipper, *46*
Orange Skipperling, *123*
Orange Sulphur, *18*

Organ Mountains, 121–22, 130–31, 133
Organ Needle, 130
Ornythion Swallowtail, *132*
Oslar's Roadside-Skipper, *30*, 31

Pacuvius Duskywing, *47*
Pahaska Skipper, *32*, 38
paintbrush, 122
Painted Crescent, *20*
Painted Lady, 22, *23*, 25
Pajarito Mountain Ski Area, 85
Palamedes Swallowtail, *138*
Pale-banded Crescent, *161*
Pale Swallowtail, *85*
Palmer's Metalmark, *121*
patrolling, 60, 65, 70. *See also* mating habits
Pearl Crescent, *65*
Pecos, 53, 79
Pecos River Valley, 65
Peloncillo Mountains, 111, 115–17
penstemon. *See* beardtongues
Percha Creek, 33, 109
Persius Duskywing, *58*, 59
Phaon Crescent, *65*
photography, 12–13
pigweed, 77
Pine Campground, 136
pines: piñon, 39–40, 99, 104, 107; ponderosa, 39–40, 42, 45–46, 50, 53, 99, 104
Pine White, *40*
Piños Altos Mountains, 49, 107
Pipevine Swallowtail, 22, *23*
Plains of San Agustín, 38
Pole Canyon, 100
Poling, Otto, 118
Poling's Giant-Skipper, *118*
Poling's Hairstreak, *133*
Polixenes Arctic, *83*
pollination, 7, 14, 119
Polydamas Swallowtail, *154*
poppies, 110
populations, 12, 25, 33
predators, 7, 9, 14, 105, 119, 136
Priest Canyon, 39
prince's plume, 95
Prop Canyon, 100
Purplish Copper, *79*
Python Skipper, *46*

Queen, *22*, 132
Queen Alexandra's Sulphur, 88, *89*, 104
Quemado, 103
Questa, 97
Question Mark, *66*

rabbitbrush, 109
railroad, 134
ratany, 131
Raton, 39, 72–73, 77
Rattlesnake Canyon Trail, 133
Rattlesnake Springs, 132
Reakirt's Blue, *19*
record keeping, 12–13, 49
Red Admiral, *20*, 21

Above:
PALE-BANDED CRESCENT

Red-bordered Satyr, *106*, 107
Red Satyr, *120*, 121
redroot buckwheat, 90, 101
Reserve, 39, 48
Rhesus Skipper, *38*
Ridings' Satyr, 38, *42*
Río Cebolla, 87
Río Chama, 87, 90
Río Grande, *97*
Río Grande Valley, 26
Río Yaqui, 117
Rita Blue, *35*
rock jasmine, 89
Rockhound State Park, 110, 124
Rocky Mountain Dotted Blue, *100*, 101
Rocky Mountain Duskywing, *41*
Rocky Mountains, *62*, 63, 72–74, 77–78, 137
Roswell, 65
Roy, 70
Ruddy Copper, 90, *91*
Ruddy Daggerwing, 116, *117*
Ruidoso, 39–40
Russet Skipperling, *90*
Rusty-tipped Page, *5*

Sachem, *22*
Sacramento Mountains, 53, 55, 57, 130, 134–37
'Sacramento' Sheridan's Hairstreak, *135*
Sagebrush Checkerspot, *106*, 107, 137
Salome Yellow, 49, *147*
Saltbush Sootywing, *36*, 93
San Juan Basin, 95, 98
San Juan River, 63, 95
San Mateo Mountains, 57
San Miguel County, 81
Sandhill Skipper, *95*
Sandía Crest, 59
Sandía Hairstreak, *34*, 35
Sandía Mountains, 35, 48
Sangre de Cristo Mountains, 53, 69, 79, 83
Santa Fe, 32–33, 39, 49, 75, 79, 93, 108
Santa Fe Baldy, 83
Santa Fe National Forest, 81
Santa Fe River Canyon, 49
Sara Orangetip, 34, *35*, 126
Sargent Wildlife Management Area, 90
Satyr Comma, 50, *51*
scree, 82
Scudder's Sulphur, *79*, 80
Seneca Creek, 72
Seven Springs Fish Hatchery, 87
shelter, 37, 60, 130, 132
Sheridan's Hairstreak, 90, *91*
Short-tailed Skipper, *41*
Siberian elm trees, 66
Sierra Blanca Mountains, 53, 130, 135
Sierra Madre Mountains, 63, 115

Above:
DINGY PURPLEWING

Sierra Madre Uplands, *62*, 63, 107
Silver-banded Hairstreak, *4*
Silver-bordered Fritillary, *86*, 87
Silver City, 35, 39, 107–09
Silver Emperor, *15*
Silver-spotted Skipper, 50, *51*, 104
Silvery Blue, *55*, 104
Silvery Checkerspot, *57*
Simius Roadside-Skipper, *38*
Sitting Bull Falls, 131
Skeleton Canyon, 111
Ski Apache ski area, 135
Slaty Roadside-Skipper, *113*
Slaughter Canyon, 132
Sleepy Duskywing, *34*
Sleepy Grass Campground, 136
Sleepy Orange, *24*
Small Checkered-Skipper, *33*
Small Wood-Nymph, *54*
Smintheus Parnassian, *81*
Smithsonian Institution, 93, 134
sneezeweed, 57, 136
Snow, Francis, 31, *49*, 108
Snow's Skipper, *104*
Soapberry Hairstreak, *69*
soapberry trees, 69
Socorro, 57
Soldier, *144*
Southern Dogface, *24*
Southern Hairstreak, *70*
Southwest Basin and Range, *61*, 106–18
Spalding's Blue, *100*, 101
specimens, 12, 108, 134
spiked gayfeather, 92
Spring Azure, *56*, *57*
Spring Canyon, 57
Spring Mountain, 81
Spring White, *34*
Square-spotted Blue, *40*
stinging nettle, 50, 58
stonecrops, 109–10
Strecker's Giant-Skipper, *69*
Striped Hairstreak, *91*
subalpine, 53–54, 74, 136
subtropical strays, *7*, 107, 114, 116–18, 128–30, 132
Subtropical Zone, 26
succession, 43, *44*, 45, 84
Sugarite Canyon State Park, 73–74, *77*
Sumner Lake State Park, 65
sunflowers, 67, 122
Sunrise Skipper, *111*
Sylvan Hairstreak, *87*

Tailed Copper, *97*
Tailed Orange, *114*
talus. *See* scree
Taos Plateau, 97
Tawny Crescent, *98*
Tawny-edged Skipper, 54, *55*, 137
Tawny Emperor, *125*
Taxiles Skipper, *50*
Tewa culture, 96
Texan Crescent, *128*
Texas, 133
Texas beargrass, 35
Texas Powdered-Skipper, *132*

Texas Roadside-Skipper, *124*
Theona Checkerspot, *122*
Thicket Hairstreak, *39*, 40
thistles, 31, 33, 38, 50, 67, 84, 87, 92, 97, 128
Three Gun Spring, 48
Tierra Amarilla, 88
Tingley, Clyde, 66
Tiny Checkerspot, *125*
Townsend, Charles, 29, 134
Trans-Pecos, 130
Transition Zone, 26–27, 29, 39–51, 53, 61, 77, 81, 95, 102
tree sap, 37, 53, 71, 125
Tres Piedras, 88
Tropical Buckeye, *128*
Tropical Checkered-Skipper, 116, *117*
Tropical Leafwing, *128*
Tropical Least Skipper, *127*
Tropical Zone, 26
Truchas peaks, 83
Tucumcari, 69
Tularosa Basin, 130, 134
tundra, 79, 82–83
Tusas Mountains, 87–88
Two-tailed Swallowtail, *33*

Uhler's Arctic, *76*, *77*
Umber Skipper, *116*
Uncas Skipper, *38*
Upper Sonoran Zone, 26–27, 29–38, 53, 61, 95
Ursine Giant-Skipper, *133*
U.S. government, 93, 139
Ute Lake State Park, 69

Valles Caldera National Preserve, 87
Variegated Fritillary, *21*, 25
velvet ash, 33
verbenas, 14, 46
vetches, 46
Viceroy, *71*, 127
Viereck, Henry, 31, 134
Viereck's Skipper, *30*, 31, 134
violets, 48, 74, 78, 136
Virden, 125
volcanic landscapes, 72–73, 77, 97, 123

Wagon Mound, 81
Wallace, Alfred, 80
Water Canyon, 46
water, 14, 37
Weed, 57
Weidemeyer's Admiral, *57*
West Coast Lady, *59*
Western Pine Elfin, *39*
Western Pygmy-Blue, *36*, 80
Western Tailed-Blue, *42*
Western Tiger Swallowtail, *56*, 57
Western White, *81*
Wheeler Expedition, 92–93
Wheeler Peak, 17, 81–83
Whipple, Amiel, 93
White Angled-Sulphur, *130*
White-barred Skipper, *110*
White Checkered-Skipper, *20*, 114
White-patched Skipper, *4*
White Peacock, *132*
White Sands National Monument, 36, 139

White-striped Longtail, 118, *119*
Whites City, 132
wild buckwheats, 35, 97, 101. *See also* specific names
wild cherries, 85
Wild Friends, 35
Willow Creek Campground, 104
willows, 79, 87, 125, 132
Woodgate, John, 86, 103, 109
Woodland Skipper, *91*
Wooton, Elmer, 134
Wooton's buckwheat, 135

Xami Hairstreak, *109*, 110

Yellow Angled-Sulphur, 116, *117*
Yucca Giant-Skipper, *34*
yuccas, 69, 133
Yuma Skipper, *97*

Zebra, 132, *133*
Zela Metalmark, *112*
Zilpa Longtail, *143*
Zuni Indian culture, 96
Zuni Mountains, 46, 86, 95, 99–103

Right:
FLORIDA WHITE

ACKNOWLEDGMENTS

This book could not have been completed without support from my New Mexico State Parks family, especially Christy Tafoya, Dave Gatterman, Tommy Mutz, and Dave Simon. Bette Brodsky, Penny Landay, Emily Drabanski, and Ethel Hess are the helpful, professional staff at *New Mexico Magazine* who brought this book to life. DeWitt Ivey's book, *Flowering Plants of New Mexico*, inspired me.

I thank my parents, John and Lorna, for encouraging independent outdoor nature exploration when I was a lad. My own family—Marcy and Seth—gracefully accepts (and so abets) my butterfly obsession.

I continue to benefit from decades of collaboration with New Mexico's Lepidoptera graybeards, notably Dick Holland, Greg Forbes, and Mike Toliver, plus enthusiastic support from a new generation including Julie McIntyre and Rachael Ryan. Unlimited collegiality from other lepidopterists, including Jim Brock, Hank and Priscilla Brodkin, Eric Metzler, Paul Opler, Mike Quinn, Bob Robbins, and Ray Stanford, is warmly acknowledged. Compadres who share my appreciation of butterflies include Nancy Daniel, Stephen Fettig, Peter Greene, Dorothy Hoard, Gerry Jacobi, Chick Keller, Randy Merker, Bruce Neville, John Pfeil, Christopher Rustay, and Linda Wiener.

I am grateful to the staff or members of the following organizations for years of fertile partnership: New Mexico Chapter of The Nature Conservancy, Native Plant Society of New Mexico, New Mexico Department of Game and Fish, U. S. National Museum, Lepidopterists' Society, North American Butterfly Association, Lincoln National Forest, Audubon New Mexico, and National Park Service.

I thank countless generous people who contributed ideas and insights during many years of shared experiences in the infinite New Mexico outdoors and ask forgiveness for any lapses on my part.

—Steven J. Cary

Right:
GREAT SOUTHERN WHITE

PHOTO CREDITS

All butterfly photographs in this book were taken by the author with the exception of the following.

Photo by Bob Barber

Page 130: Orange-barred Sulphur

Photos by Jim Brock

Page 94: Mohave Sootywing
Page 98: Great Basin Wood-Nymph
Page 111: Sunrise Skipper
Page 114: Arizona Metalmark
Page 117: Yellow Angled-Sulphur
Page 129: Hammock Skipper
Page 132: Ornythion Swallowtail
Page 133: Ursine Giant-Skipper
Page 162: Dingy Purplewing
Page 165: Florida White

Photos by Hank & Priscilla Brodkin

Page 1: Gray Cracker
Page 3: Isabella's Heliconian
Page 5: Drusius Cloudywing
Page 5: Rusty-tipped Page
Page 15: Silver Emperor
Page 94: Northern White-Skipper
Page 113: Moon-marked Skipper
Page 117: Ruddy Daggerwing
Page 118: Nabokov's Satyr
Page 161: Pale-banded Crescent

Photos by Jeff Glassberg

Page 100: Rocky Mountain Dotted-Blue
Page 141: Brazilian Skipper
Page 147: Salome Yellow
Page 157: Broad-banded Swallowtail

Photo by Axhel Muñoz

Page 121: Palmer's Metalmark

Photos by Paul Opler

Page 88: Freija Fritillary
Page 98: Tawny Crescent

Photo by Jane Ruffin

Page 11: Fulvia Checkerspot ovipositing

Photo by Jennifer Sanders

Page 158: Crimson Patch

Photo by Bob Sivinski

Page 130: White Angled-Sulphur

Photo by Mark Watson

Page 12: Blackened Bluewing

Photo by Dale Zimmerman

Page 48: Nokomis Fritillary